30.00
YALE 060390
1255710V
EGGLASY-1
25719 M

A History of the Ecosystem Concept in Ecology

A HISTORY OF THE ECOSYSTEM CONCEPT IN ECOLOGY

More Than the Sum of the Parts

Frank Benjamin Golley

Yale University Press

New Haven and London

Published with assistance from the foundation established in memory of Philip Hamilton McMillan of the Class of 1894, Yale College.

Designed by Deborah Dutton.
Set in Galliard Text and Gill Sans Display types by Keystone Typesetting, Inc., Orwigsburg, Pennsylvania.
Printed in the United States of America by Vail-Ballou Press, Binghamton, New York.

Library of Congress Cataloging-in-Publication Data
Golley, Frank B.
    A history of the ecosystem concept in ecology : more than the sum of the parts / Frank Benjamin Golley.
        p. cm.
    Includes bibliographical references and index.
    ISBN 0-300-05546-3
    1. Ecology—History. 2. Biotic communities—History. I. Title.
QH540.8.G64   1993
574.5'09—dc20   93-17577
                                                                                      CIP
A catalogue record for this book is available from the British Library.

The paper in this book meets the guidelines for permanence and durability of the Committee on Production Guidelines for Book Longevity of the Council on Library Resources.

10 9 8 7 6 5 4 3 2 1

To Priscilla McKinzie Golley

CONTENTS

ABBREVIATIONS

AEC	Atomic Energy Commission
BES	British Ecological Society
CNRS	Centre National de la Recherché Scientifique
EDFB	Eastern Deciduous Forest Biome project
ELM	Grassland system model
ENCORE	European network of catchments for ecological research
FAO	Food and Agriculture Organization
IBP	International Biological Program
ICSU	International Council of Scientific Unions
IGY	International Geophysical Year
IUBS	Union of Biological Sciences
IUCN	International Union for the Conservation of Nature
MAB	Man and the Biosphere program
MEDECO	Mediterranean Ecological Society
NAS	National Academy of Science
NSF	National Science Foundation
PERT	program evaluation and review technique *(sys. eng.)*
RANN	Research Applied to National Needs
RES	Regional Environmental Systems
SIL	*Societas Internationalis Limnologiae Theoreticae et Applicatae*

SREL	Savannah River Ecology Laboratory
UNESCO	United Nations Educational, Scientific, and Cultural Organization
WHO	World Health Organization

PREFACE

I was motivated to write this book for a variety of reasons. First, I intended to write a critical review of the ecosystem concept. All phases of science need critical review and ecology is no exception. Review is a bit like checking one's bearings when on a hike in the mountains. It is important to know where you are. The ecosystem has been a key concept in the development of modern ecology, yet today it is widely misunderstood and misused. Many of my students in the 1980s came to graduate school thinking the ecosystem concept was old-fashioned and that it had been rejected by ecologists. It was clear that ecologists did not share a common understanding of the concept, nor were they agreed on how ecology should be taught to the beginning student. Part of the reason for this situation is the information explosion in ecology. As I collected data for the review of the ecosystem concept, I became overwhelmed by the quantity of information. In seeking to organize it, gradually I became convinced that a chronological organization would be more useful than the traditional division into structure and function.

Therefore, the project evolved from a scientific review into a history of the ecosystem concept. Ecology as a subject is little more than one hundred years old, and recently it has experienced a period of rapid growth in public awareness. I have found few students, and few people, who have a clear understanding of the development of scientific ecology. In fact, the time frame for analysis of a topic or a concept in a graduate ecology course seems to have a depth of only five to ten years. Although this limited perspective reflects the amount of information students can assimilate in any subdiscipline of ecology, it also reflects the fact that there are few historical studies of ecology: it is taught as if it had no history. Thus the historical approach is distinctly useful beyond the focus on an important concept.

It is obvious that the review of a scientific concept and its history represent two very different perspectives. The study of history is a discipline that has its own methods, canons of scholarship, and standards. Contrary to what many scientists might imagine, a history is not merely a chronological listing of past events. Rather, the events must be examined, questioned, and placed in an appropriate context. Studying the history of ecology is like peeling an onion. Once one layer is removed, a new layer appears, so causation and motivation are linked, layer upon layer, in time. Further, the interaction of the historian with these interpretations changes as the story becomes more and more complex. There is no general history of ecology, no set of facts that represents ecological change in time. Instead, each event is as unique as each individual organism is to the field ecologist; each historical interpretation is as limited as each ecological hypothesis; and each historian is as limited by his or her experience and capacity to understand as each ecologist is in beginning a new research project in a new habitat.

I can cite an example of this phenomenon, associated with this particular story. In September 1992, Joel Hagen published his history of the ecosystem concept in *An Entangled Bank*. Hagen and I use approximately the same material, but we each tell two very different stories about ecosystem ecology. His approach is to embed the ecosystem concept in a history of American and English ecology. My approach has been to focus on the concept and the individuals contributing to concept development and then to consider causal factors that led to the pattern we observe. Joel Hagen is a historian and comes to the story from outside, as it were; I was a participant in the story and approach it from within. The result is two different books.

All disciplines have standards by which the usefulness of their creations are judged. I would apply two criteria. The first is generality. Does the study open the window to larger vistas? Does it provide links to other events occurring at the same time? Does it link our present concerns with those of ecologists in other times and places? The second is uniqueness. Does it possess a singular voice so that the individual behind the interpretation becomes visible? The voice of Fernand Braudel, the late master of French history, is distinct and clear even in translation. I can make the same comment about contemporary environmental historians such as Bill Cronin, Al Crosby, and Donald Worster. I hope that in this book I have satisfied these criteria.

As I proceeded in writing this book, I realized more and more that my own role in the development of the ecosystem concept and in the many activities that created the concept as we know and use it today influenced my perspectives and interpretations. My view has been that of a participant, not merely of an observer. I have tried to check the tendency to rely on memory and express personal convictions, yet inevitably at times the personal voice intrudes. For this reason, I provide more biographical comment than is usual in a preface.

My particular approach to this enterprise comes partly from my family background and partly from my graduate training. My maternal grandfather, Lewis C. Baird, was a historian, army officer, and engineer. He wrote several personal histories, as well as the *History of Clark County Indiana,* published in 1909, which was one of many county histories of that era. I was raised with my grandfather's histories as models of family and public life. As a graduate student at Washington State University when I began a master's program in zoology and wildlife management, Herbert Eastlick was chair of the department of zoology and Helmut Buechner was my major professor. These men taught that a scholar should know his or her subject critically and that the library was as important to the ecologist as the laboratory or field. I was required to make a bibliography of the literature pertaining to my research topic, to read all of the papers and books in this literature, and to incorporate them into my thesis. Each evening I carried a shoe box that contained my bibliography on three-by-five cards to the library. Thus I became attracted to the bibliographic review and thorough analysis of a topic from the beginning of my scientific career.

My graduate training continued at Michigan State University, where I was fortunate to be guided into a study of the energetics of a food chain by Don W. Hayne. This study was one of the first of its kind, and through it I met Eugene P. Odum and became associated with him at the University of Georgia, where I have remained throughout my career, working on ecosystem studies, among other projects, and have been involved in many of the major ecosystem efforts nationally and internationally. These efforts included the International Biological Program, the ecosystem studies of the U.S. Atomic Energy Commission, the UNESCO Man and Biosphere projects, and most recently work with the organization of the Association of Ecosystem Research Centers. I served as division director of the National Science Foundation Division of Environmental Biology (1979–81), which provides financial support to ecosystem studies.

One can, of course, justify an interest in reviewing a concept on other than personal grounds. In 1964, the biologist Bentley Glass wrote in defense of reviewing in the *Quarterly Review of Biology.* Glass quoted Robert Graves: "The scholar is a quarryman, not a builder, and all that is required of him is that he should quarry cleanly. He is the poet's insurance against factual error." Glass goes on to argue with Graves:

But I would say that the real scientist, if not the scholar in general, is no quarryman, but is precisely and exactly a builder—a builder of facts and observations into conceptual schemes and intellectual models that attempt to represent the realities of nature.

This insight, this vision of the whole of nature—or at least some larger part of it—exists in all degrees among the individuals we call scientists. The man who adds his bits of fact to the total of knowledge has a useful and necessary function. But who would deny that a role by far the greater is played by the original thinker and critic who discerns the broader outlines of the plan, who synthesizes from existing knowledge through detection of the false and illumination of the true relationships of things a theory, a conceptual model, or a hypothesis capable of test?

The creativity of scientific writing lies precisely here. The task of the writer of a critical review and synthesis that fulfills these objectives and meets these criteria is not only indispensable to scientific advance, it surely constitutes the essence of the scientific endeavor to be no mere quarryman but in some measure a creator of truth and understanding. The aesthetic element that makes scientist akin to poet and artist is expressed primarily in this broader activity.

In this examination of the development of ecosystem studies, I also have been guided by a model from Arthur Lovejoy, who wrote *The Great Chain of Being*. Lovejoy subtitled this book "The History of an Idea," and his phraseology captured my imagination. The ecosystem is an idea, a powerful idea, and a popular idea. I have tried to keep this always in mind. Later, I learned that the study of the history of ideas is a subdiscipline of history. I do not pretend to have gone that far. Rather, I use the phrase metaphorically to describe my intent and goal.

Finally, I apologize for not being encyclopedic. There is an advantage to such an aim, and my first intention was to include everything pertinent to the story within this volume. But scientific literature is not a linear stream of papers and books, each building on the other. Rather, it is a complex of forward motions, turnings, repeating, and contradictions. Individuals enter for a time, are active, and then turn to other pursuits. Therefore, instead of an encyclopedia, I have adopted the metaphor of the drama, with a stage, principal characters, and a cast of hundreds. Unfortunately, because of space and time restrictions, I have had to exclude many of those standing immediately offstage. The script moves through time in a straightforward fashion, but each act has a separate identity.

What do we now have? We have elements of a review, a history, a drama, and certainly, a story. Bill Cronin, at a Duke University gathering of environmental historians, said that history is storytelling. He emphasized that since we have a story, we should be aware of a teller of the tale, a store of past events from which the teller selects, an audience, and an environment that encloses the storyteller and his audience. I hope that each of these specific elements is clear in what lies ahead.

ACKNOWLEDGMENTS

As in any venture of this kind, an author has many collaborators and assistants. I am no exception. Two individuals have been especially important. First, Eugene Odum generously spent many hours and lunches discussing his own perspective of the history of the ecosystem concept. His insights, and especially his memories of those processes contributing to his own contributions to ecology, were unique and invaluable to me. Second, my wife, Priscilla M. Golley, read, edited, and commented on the manuscript throughout its development, and I am indebted to her for this direct assistance, in addition to providing me with an environment in which to think and write about ecosystems. I dedicate this book to her. Among the many others who have provided me assistance in various ways are Virginia Benjamin, F. Herbert Bormann, Per Brink, Jerry Brown, Robert Burgess, Tom Callahan, David Coleman, Betty Jean Craig, Susan Curtis, Francesco di Castri, John Edwards, Frederick Ferré, Peter Grubb, Wolfgang Haber, Arthur Hasler, Gilbert Head, Bengt-Owe Jansson, Carl Jordan, Sven Jorgensen, Hiroya Kawanabe, Bengt Landholm, Joseph D. Laufersweiler, Gene Likens, Orie Loucks, William Loughner, Helen MacCammon, Mark McDonald, James MacMahon, John Magnuson, Ramon Margalef, Albert Meier, Florencia Montagnini, Makato Numata, H. T. Odum, Wilson Page, Bernard Patten, Charles Reif, William Reiners, Hermann Remmart, Ann Richards, Paul Richards, Thelma Richardson, Lucy Rowland, Lech Ryszkowski, Emily Russell, Vincent Schultz, John Sheail, Harald Sioli, Mart Stewart, Herman Sukopp, Barbara Taylor, Gerhard Trommer, Richard West, and Mark Westoby: I am grateful to them all for their assistance. I also express thanks to the reference staff of the University of Georgia Science Library, to the staff of the Archives and Records Management Department, University of Georgia Library, to the library of the Botany School, Cambridge University, the American Heritage Center the University of Wyoming Library, the President and Fellows of Magdalen College, Oxford, and to Sterling Library, Yale University, for giving me access to the papers and photographs in its collections. I am grateful to the University of Georgia Institute of Ecology, which provided me support to complete this project. I also acknowledge with appreciation permission to reprint materials from the publishers, institutions, and individuals listed below:

Academic Press; the American Association for the Advancement of Science; *American Midland Naturalist; American Scientist;* the Argonne National Laboratory; A. P. Watt, Ltd.; Bernard Patten; Blackwell Scientific Publications; Cambridge University Press; the Carnegie Institution of Washington, D.C.; Charles B. Reif, Wilkes-Barre, Pennsylvania; Chapman and Hall; Dover Publications; the Ecological Society of America; Indiana University Press; the Institute of Ecology, University of Georgia; *Limnology and Oceanography;* Macmillan Publishing; the President and Fellows of Magdalen College Oxford; Oxford University Press; Pitman Publishing; Springer-Verlag; Prof. T.ap. Rees, University of Cambridge, Department of Plant Science; University of Chicago Press; E. P. Odum Papers, Department of Archives and Records Management, University of Georgia; Raymond Laurel Lindeman Papers, Manuscripts and Archives Collection, Yale University Library; Yale University Press; and W. B. Saunders.

Frank B. Golley
Athens, Georgia

A History of the Ecosystem Concept in Ecology

CHAPTER I

Introduction

This story concerns the development of a scientific concept—the *ecosystem*. Metaphorically speaking, its development has a life history. There was an exact moment of birth: when the English ecologist Arthur Tansley created the word and presented it in a technical paper. Seven years later the term was applied in one of the most significant studies in modern ecology, the study of the trophic dynamics of Cedar Bog Lake, Minnesota, by Raymond Lindeman. The Lindeman work ushered in a period of growing interest in ecosystem studies. Eugene P. Odum's use of the ecosystem concept as an organizing concept, however, in his popular, widely used textbook, *Fundamentals of Ecology,* transformed a specialized technical idea into a concept with vast theoretical and applied significance.

Following the publication of Odum's textbook, ecosystem studies increased rapidly, as did its theoretical development. The ecosystem concept became a guiding paradigm of ecological science. The process of development culminated in the institutionalization of ecosystem studies in the International Biological Program (IBP). In the United States at least, IBP was dominated by the ecosystem approach. Ecological modeling was intended to integrate the data collected by teams of scientists working at research sites that were considered typical of wide regions. The IBP studies were defended before hearings of the U.S. Congress, where ecosystem studies were interpreted for lawmakers and their applied significance described as important and needed.[1]

The IBP never achieved the goals and objectives contemplated by its organizers and boosters. It did, however, have significant scientific success, and it led to permanent funding for ecosystem studies within the U.S. National Science Foundation (NSF). This process of institutionalization created the conditions for a maturing of ecosystem studies, a period in which descriptive studies and experiments could be designed and carried out in a logical fashion devoid of the hurly-burly of the adolescent growth period. An example of a successful strategy of research in this final period was that developed and applied by F. Herbert Bormann and Gene E. Likens at their Hubbard Brook, New Hampshire, research site. Likens and Bormann studied the ecosystem as an object and then manipulated the object to discover how it was constructed, how it functioned, and how it responded to disturbance and stress. These Hubbard Brook studies were the models for many other ecosystem programs and became a source of data for environmental management.

The ecosystem story is largely an American tale. Although its genesis and birth were partly in Europe—where ecology became self-conscious when Ernst Haeckel coined the term *ecology* in 1866[2] and many ecosystem-like efforts were subsequently initiated—the conditions for growth existed only in the United States. The most important factor was the second world war, which interrupted ecological work worldwide. After the war the United States experienced a period of rapid development, which included scientific activity in ecology. In contrast, Europe and Japan were preoccupied with reconstruction, a reexamination of the principles of government, and the recovery of normal life. Not only were many of the ecologists in Europe who had played or could play a role in ecosystem studies dead or elderly, but ecologists also repudiated those aspects of ecological theory that had been used by the Nazis and militarists to force conformity on the population and to base racist policy. Ecosystem studies were too close to prewar organismic theories of ecological and social organization to be popular.

In America, however, the ecosystem concept appeared to be modern and up to date. It concerned systems, involved information theory, and used computers and modeling. In short, it was a machine theory applied to nature. The concept promised an understanding of complex systems and explicitly promised to show how Americans could manage their environment through an understanding of the structure and function of ecological systems and by predicting their responses to disturbance. Further, it extended the holistic concept into the modern, postwar environment. The whole had played a key role in the intellectual development of ecology and in the United States served as an organizing paradigm of prewar ecology through the monoclimax and successional paradigms of Frederic Clements, the famous plant ecologist of the Carnegie Institution of Washington. The concept of holism had wider cultural significance. It postulated the existence of a complex entity, larger than humans or human society, which was self-organized and self-regulating. In one sense, the whole was an extension of the Mother Earth idea in modern guise. It involved the extension of God-like or parental properties to nature. Most significantly, it provided the individual faced with the complications and difficulties of daily life the notion that somewhere out there, there was ultimate order, balance, equilibrium, and a rational and logical system of relations. This mixture of ideas was carried forward past the second world war period by the generation that had fought the war, and it dominated the immediate postwar years. The ecosystem concept fit into it, giving guidance to ecological scientists and avoiding dissonance with the overall culture.

Ecosystem ecologists benefited from close involvement in the concerns of American society. The major source of their funding was initially the U.S. Atomic Energy Commission (AEC). Ecologists benefited from the cold war and the Korean War as they studied the base-line ecology of nuclear weapons facilities, the effects of radiation on organisms, and the movement of radioactive materials through food chains.

By the end the 1950s, however, the conflicts within American society began to intrude upon ecological scientists, no matter how deeply they hid within their ivory towers. Rachel Carson ignited the environmental movement through her book on the effects of pesticides. Ecologists were asked to testify on both sides of the debates that followed. Environmentalists seized upon the ecosystem concept as a way to maintain their faith in holism. The use of pesticides by humans disturbed in a fundamental way the natural order of the world. The issue was a moral one. The ecosystem, and sometimes "the ecology," were being disturbed, and humans were in danger of destroying the system upon which they lived. The manager and industrialist found the ecosystem equally attractive. It promised a way to manage complex natural systems. With adequate understanding of the ecosystem we might use salt marshes to process sewage and industrial wastes. Knowledge of forest ecosystems would help us to use forest management practices that would yield an optimum product with minimal damage to streams and soils. The ecologist, like the proverbial spider in the center of its web, passively accepted the buzzing activity, benefiting from the attention and the resources generated by the environmental crisis. Students flocked to ecology classes, the numbers of ecologists increased rapidly, the Ecological Society of America struggled with the identification of ecologists as different from environmentally oriented engineers and physicians, and funding expanded for research, study, service and institutions. The growth of ecosystem studies occurred in this environment.

In this period of hectic activity, theoretical contradictions and methodological inconsistencies were noted and then overlooked. However, as ecologists were called upon to solve serious problems involving complex ecosystems, these technical shortcomings became more evident. The complex systems, sometimes consisting of thousands of species, were simplified by the theoreticians to sometimes as few as three or four components. The myriad environmental interactions between ecosystem components were reduced to a few flows of energy. The dynamic response of natural systems was made deterministic, being consistent with physical theory. The ecosystem was conceived as a machine, represented as a computer model. In IBP, this conception was pushed to its logical extreme by George Van Dyne and his colleagues in their study of grasslands. The study was not successful. Most ecologists turned away from the study of the ecosystem and focused instead on the processes or parts of ecosystems, claiming that it was necessary to understand the parts before one could understand the whole. Even the NSF ecosystem studies program was captured by the process approach for a time.

Yet those ecologists interested in the ecosystem returned to an idea pioneered in Europe in the early part of the twentieth century. This approach involved the study of the whole system directly. The ecosystem was perceived as an object, and all the usual scientific approaches and tools were applied to understand the object and its behavior. In the prewar ecological tradition, these objects were almost always lakes, because on land the boundaries of the ecosystem object were less clear. Bormann and Likens solved this problem of identifying the boundaries of terrestrial ecosystems by identifying the watershed as the object of interest. The watershed divide, especially in mountainous terrain, was precise. Water flows into the watershed from rainfall and snow, it accumulates in the soil and rock, and if the watershed is hydrologically sealed at the bottom by suitable geological strata, all the water leaves through the stream, which exits at the bottom or through evaporation and transpiration. If the ecologist measures the input through precipitation and the output through the stream, then the system dynamics of water, water chemistry, and water biology can be studied directly. Further, the watershed can be manipulated by cutting the trees or burning the understory shrubs, and the impact of the experiment can be measured. Thus, the watershed approach provided a way to set ecosystem studies on an experimental foundation. Its current success and utility stems from this modification.

In addition to the problems of applying the machine metaphor to complex natural systems, ecosystem ecologists in the growth phase of the subject were faced with a challenge within the field of ecology. In the late 1950s, associated with the centennial of Darwin's publication of *The Origin of Species*, ecologists reexamined the role of evolution in the development of behavior and life history, the regulation of populations, and the organization of communities. One stimulus for this reexamination was V. C. Wynne-Edwards publication of a book in which he argued that selection may occur at the level of the group. Wynne-Edwards was interested in colonial birds, and he drew some of his examples from this level of ecological organization. Wynne-Edwards's book threw the question of the level at which selection operated into sharp relief. A strong and hostile response to group selection occurred. Emphasis was focused on selection at the level of the individual organism. John Harper, the British plant ecologist, said:

> Ultimately all the discoveries of descriptive, production or population ecology must find their meaning in evolutionary phenomena. . . . Evolutionary thinking concentrates attention on the behavior of the individual and his descendants. If nothing in biology has meaning except in the light of evolution *and* if evolution is about individuals and their descendants—i.e. fitness—we should not expect to reach any depth of understanding from studies that are based at the level of the superindividual. . . . What we see as the organized behavior of systems is the result of the fate of individuals.[3]

This assertion was a serious challenge to ecosystem ecology. It was ignored.

The evolutionary ecology challenge was ignored by ecosystem ecologists initially because it was framed in terms that were not relevant to their interests. They had relatively little interest in species because they collapsed species into categories such as trophic levels. Further, ecosystems were being investigated with almost no time-space relations. The problem was to capture enough real system behavior at a given point in time and space to represent the system. For this work, the theory of evolution was of little direct interest in the interpretation of ecosystem data.

The impact of evolutionary ecology on the ecosystem paradigm began to have an effect as competitors to Odum's textbook appeared and as students began to be trained in this new ecology. Thus, coincident with the rapid increase in the numbers of ecology students was the presentation of an alternative paradigm of ecology. Further, evolutionary ecologists asserted that scientific purity could only be achieved within their approach. Harper's words reflect this demand for orthodoxy. As evolutionary ecologists obtained control of ecology journals their views began to change the editorial policy of the mainstream publications. Gradually, new journals that continued or applied the ecosystem theme were created by new ecological societies and commercial publishers.

A key element in the competition between ecosystem studies and evolutionary ecology was the role of the biology department in the university. Universities are bastions of conservatism, dedicated to teaching the young current knowledge in a linear and predictable sequence. Biology, fractured into botany, zoology, and other taxonomic or functional "ologies," also was evolving during this period of transformation and growth in ecology. Contrasted to the great discoveries in molecular and cellular biology, especially those associated with DNA and the chromosome, were the conservative fields of anatomy, physiology, taxonomy, and ecology. With limited resources in universities and much larger resources in the federal government, especially of the United States, molecular biology and biochemistry were able to overcome the hold traditional subjects had on biology and establish control in many biology departments, even occasionally ousting ecology and natural history studies entirely.

These processes of change were significant because ecology had developed within biology, although in theory it was the study of the interaction of living systems and the environment. The environment was seldom accorded the same level of attention by ecologists as the biological side of the duality. Ecosystem ecology did not fit into the academic biology department at all well. Ecosystem study required teams of scientists, and in the university individual scholarly work was held in highest esteem. It focused on complex systems, which were often beyond the technical capability of a single investigator to study in more than a superficial manner. Yet doctoral students, the work horses of university research, could not easily undertake the study of an ecosystem as a thesis topic. The facilities and logistic support needed for ecosystem studies were difficult to fit into the academic environment, where facilities were rewarded to active units, growing rapidly or commanding power and prestige.

As a consequence, much ecosystem research developed in new institutions inside and outside universities. The national laboratories of AEC were the first places to provide a home for these researchers, but gradually institutes and centers were formed that shared in supporting ecosystem studies. Nevertheless, the nature of the ecosystem as a new object for scientific study, justifying a new scientific field, was never fully realized. Only in 1988 in the United States was an organization established to focus exclusively on the needs of ecosystem science.[4] It has no parallel elsewhere.

The story of the ecosystem concept involves scientific personalities, the dynamics of practice and theory in science, interaction and competition among the sciences, the influences of institutions and societies, the role of great events such as wars, depressions, and revolutions, and the role of cultural paradigms, which structure how we think about nature and our expectations of science. In telling the story of the ecosystem concept I adopt a chronological approach, beginning with the birth of the term in chapter 2. In chapter 3 I examine the background of the ecosystem approach the first application of the concept in studying a real ecosystem. In chapter 4 I focus on the period of transition from technical term to dominant scientific paradigm. The chapter begins with the publication of Odum's textbook and ends in the mid-1960s with a discussion of the theoretical and technical developments of ecosystem studies. These studies then were transformed into well-funded, active, large programs of research and training through the International Biological Program. IBP is the focus of chapter 5, with emphasis on the United States biome programs, which most clearly represented ecosystem studies in this period. There were, however, other IBP ecosystem programs in the United States and other countries. These nonbiome efforts are the subject of chapter 6. The major point of chapter 6 is that within one ecosystem program, the Hubbard Brook program, a workable scientific approach for the study of the ecosystem was clearly demonstrated. This approach led into the mature phase of the science and continues as the dominant form of ecosystem research today. The story ends at about 1975. In chapter 7 I reexamine the ecosystem story, pointing out key steps in its development, and draw conclusions about the development of this scientific concept in a technical, philosophical and cultural context.

CHAPTER 2

The Genesis of a Concept

In 1935 Alfred George Tansley (1871–1955) introduced a new word to the world. *Ecosystem* referred to a holistic and integrative ecological concept that combined living organisms and the physical environment into a system. Tansley presented the ecosystem concept in a twenty-three-page article titled, "The Use and Abuse of Vegetational Concepts and Terms" in the scientific magazine *Ecology:*

> But the more fundamental conception is, as it seems to me, the whole *system* (in the sense of physics), including not only the organism-complex, but also the whole complex of physical factors forming what we call the environment of the biome—the habitat factors in the widest sense.
>
> It is the systems so formed which, from the point of view of the ecologist, are the basic units of nature on the face of the earth.
>
> These *ecosystems,* as we may call them, are of the most various kinds and sizes. They form one category of the multitudinous physical systems of the universe, which range from the universe as a whole down to the atom. (Tansley, 1935, 299)

Thus, Tansley's ecosystem concept identified a system that was: (1) an element in a hierarchy of physical systems from the universe to the atom, (2) the basic system of ecology, and (3) composed of both the organism-complex and the physical-environmental complex. From this origin in 1935 to the present, Tansley's ecosystem has remained a key concept in the ecological sciences.

TANSLEY THE MAN

Certainly the success of the ecosystem concept was partly due to the distinguished reputation of its creator. Arthur George Tansley was born in 1871 in London, the only son of a businessman, George Tansley, who had retired at an early age to give his time to voluntary teaching and public work. Arthur George Tansley was attracted to the study of field botany and entered the University of London to attend lectures in the biological sciences before completing his preparatory studies. He went to Trinity College, Cambridge, in 1890, where he read botany, zoology, physiology, and, in the last year, geology. He received a first class in part 1 of the Cambridge Natural Science Tripos, whereupon he returned to University College, London. There he served as assistant to the distinguished botanist F. W. Oliver. Tansley stayed in London until 1906, during which time he repeated his performance in the second part of the Cambridge Natural Science Tripos. He then returned to Cambridge as a lecturer at the Cambridge Botany School under Prof. A. C. Seward.[1]

Although Tansley began his scientific work as an anatomist, he was directed toward plant ecology through his early interest in field botany. Later, as a trained scientist, he was influenced by a trip he made to Ceylon and the Malaya Peninsula in 1900/01 and by Oliver's interest in field studies of maritime communities. He had initiated the scientific journal *The New Phytologist* in 1902, which he edited for thirty years (1902–31). Through this journal Tansley was able to provide an avenue for research that did not fit the conservative standards of the botanical journals. This research included ecological studies. In 1904 he called for establishment of a vegetation survey of the British Isles, and in 1912, when this survey had lost its momentum, he worked to form an association of ecologists. In 1913 the first ecological society in the world was organized, with Tansley as its president. He was made a fellow of the Royal Society in 1915 and appointed to the Sherardian Chair of Botany in Oxford in 1927 (fig. 2.1), in which position he served until his retirement in 1937. He was knighted in 1950.

The biological details of Alfred George Tansley suggest that he was not only an acknowledged leader of British plant ecology but that he achieved substantial personal success as a result. Actually, the situation was quite different. Tansley struggled throughout his career to open traditional botany to ecology and to have ecology equally accepted as part of the natural sciences. His

2.1 Painting of Sir Arthur George Tansley when he served as professor of botany, Oxford University. Reproduced with permission of the President and Fellows of Magdalen College, Oxford

commitment to service in his profession, together with tolerance and persistence, eventually led to success. His professional advancement was delayed, however, and the frustration this delay caused must have been intense.[2] In 1912 he contemplated moving to an academic post in Australia, and in 1922—partly from his concern for first world war veterans suffering mental problems from trench warfare and partly from his frustration with academic life—he visited Sigmund Freud in Vienna to further his study of psychology. The following year he resigned his lectureship and moved his family to Vienna to study with Freud[3] and initiated a practice in psychology when he returned to Cambridge. Although his practice was soon abandoned, his textbook *The New Psychology and Its Relation to Life,* was successful and was reprinted and translated several times.

Arthur Tansley was a highly motivated, effective leader for ecologists, with a deep interest in philosophy and psychology, personally familiar with the pioneers of ecology in other countries, and a scientist with exceptionally high standards. Chiefly through his work as an editor, Tansley set a standard for ecological science, and it was from his complex personality that the ecosystem concept emerged.

Terms and Concepts of Vegetation

Tansley's article on vegetational terms and concepts was an invited contribution for a festschrift for another ecologist, Henry Chandler Cowles (1869–1939). Cowles was associated with the University of Chicago and at the turn of the century carried out a series of studies on plant succession on the sand dunes of Lake Michigan, as well as other habitats in the Chicago area (Cowles, 1899, 1901). Tansley offered his article as a contribution to a subject in which Cowles had special interest. He honored Cowles by saying, "During the first decade of this century indeed Cowles did far more than any one else to create and to increase our knowledge of succession and to deduce its general laws."[4]

The immediate stimulation for the subject of the treatise on vegetational terms and concepts, however, was, according to Tansley, the appearance of four articles by the South African ecologist John Phillips,[5] which concerned the biotic community, succession, development, the climax, and the complex organism (Phillips, 1931, 1934, 1935a, 1935b). In these articles on succession and community organization Phillips related the concepts of the American ecologist Frederic Clements (1874–1945) to philosophical concepts of the biotic community as a complex organism and as a philosophical whole, after the ideas of Jan Christian Smuts (1870–1950). Phillips (1931) devoted most of his first treatise on the biotic community to his argument that ecologists should consider both plants and animals as members of a biotic community. Reading before the 1930 International Botanical Congress, Cambridge, he emphasized the place of animals in the organization and structure of communities.

> When elephant [sic] frequent any portion of forest for any length of time, they are invariably followed by the scavenging *Potamochoerus choeropotamus* (wild pig), and at times by baboon, which take advantage of the roots and bulbs displaced by the great animals, and which are not above searching the droppings for food. The disturbance to the soil caused by the elephant, the wild pig and the baboon, brings about soil improvement, and stimulates many dormant seeds to germinate. Fruits passed through the animals are cleaned of their outer coverings and fall into improved germinating beds. Naturally a certain proportion of the fruits is spoilt in the process of passing through the animals, while existing regeneration may be destroyed. (Phillips, 1931, 11)

Because ecologists tended to restrict their attention to plants or animals exclusively, Phillips was reiterating an important point made earlier by A. G. Vestal (Illinois plant ecologist) (1914), Frederic Clements (1916), and Victor Shelford, animal ecologist at the University of Illinois (1926).It was only in his conclusion that Phillips departed from the evidence:

> It is in keeping with the importance of the subject that I should at this juncture refer to a further aspect of the community—that aspect that has already called for criticism from certain quarters—the community as a *complex organism*. Clements (8, p. 199; 10, p. 3; 16, p. 314) in his purpose of introducing the term and view appears to have been misunderstood by some (18, 24), but has had the support of Tansley (41, p. 123; 43, p. 678), provided the term *quasi*-organism is employed and provided the concept applies to Tansley's *autogenic* succession. Briefly Clement's purpose is to emphasize the organic entity of the community, his epithet *complex* immediately distinguishing this *communal* organism from the *individual* organism of general terminology. While I—and doubtless Clements himself—would agree that philosophically General Smuts (40, pp. 339–43) by his masterly and inspiring exposition—in a universal connection—that groups, societies, nations, and Nature are *organic without being organisms,* are holistic without being wholes—has pointed to the truth, I still am able to see that the concept of the *complex organism* has much to commend it in practice. It certainly focuses attention—and such a focussing is essential to advance—upon the place and function of all life in that organic entity the community.
>
> A biotic community in many respects behaves as a complex organism—in its origin, growth, development, common response, common reaction, and its reproduction. In accordance with the holistic concept of Smuts (1926), the biotic community is something more than the mere sum of its parts; it possesses a special identity—it is indeed a mass-entity with a destiny peculiar to itself.[6]

Phillips's other three articles each presented an aspect of a single argument and expanded upon his ideas (Phillips, 1934, 1935a, 1935b). His purpose was "a careful review of the highly polemical field concerned with the essential nature and the direction of *succession* and *development*; the nature of the *climax*; the existence of the *complex organism*; and the inherent oneness of the *complex organism* and the *biotic community*" (Phillips, 1934, 555). Phillips's technique in these articles was to pose a series of questions concerning the many possible interpretations of a concept, review the literature—showing how various ecologists addressed the questions—and then to draw his own conclusions. He derived his answers apparently entirely arbitrarily, but with reference to the authority of Clements, Smuts, Tansley, and others.

The intellectual idea Phillips was advancing in his articles affected a major topic of ecological discussion and argument: the nature of the biotic community. The creative element in Phillips's writings was the connection of the thought of Clements and Smuts in defense of Clements's concept of the community as a complex organism. In the first part of a three-part series, Phillips concluded, with Clements, that ecological succession is always the result of biotic reactions on the environment and that it is always progressive. That is, succession is convergence toward an end-point, called the climax community, and it represents a process of development of a complex organism.

In his second treatise he dealt with two aspects of the Clementsian paradigm. First, he asserted that the process of development causes integration to occur among the biota within the community. Second, the climatic climax is set by the regional climate, and there is ideally only a single climax (the monoclimax) in a region. That is, the climax community is in a dynamic equilibrium with the climax habitat, but this equilibrium is not static or permanent.

Phillips's final treatise was a philosophical defense of the concepts of emergence and the complex organism. Phillips thought that emergence "appears to offer [the ecologist] a vantage point from which to survey characteristics of novelty, of integration, of wholeness, emergent from succession and development in biotic communities. . . . Communities are not mere summations of individual organisms, but are integrated wholes with particular emergents" (Phillips, 1935b, 490).

Phillips did not present new evidence for these community concepts. Rather, he was reviewing the arguments about the origin and character of ecological communities and defending the Clementsian position. Part of his strategy in these articles was to appeal to an authority to make his argument more convincing. Frederic Clements was an authority figure in American ecology, and his theories of plant succession and the nature of the community were widely accepted by American ecologists. According to Phillips, Clements provided the most inclusive explanation of these phenomena. Jan Christian Smuts was the prime minister of South Africa and an authority figure for Phillips. In turn, Phillips also involved Tansley, the most distinguished British ecologist, as a further but more tangential authority for his integrating treatises.

The relation between Phillips, Clements, Smuts, and Tansley is complex. Phillips was younger than the other three. He was personally acquainted with Clements, having visited him in America. He was attracted to Clements's ideas, had introduced Smuts's ideas to Clements, and became an advocate of Clementsian ecology in Africa. Phillips was also personally connected with Smuts,

2.2 Participants in the International Phytogeographical Excursion, in England, 1911. Seated on the ground, left to right: T. W. Woodhead, C. E. Moss, Frederic Clements, Weiss; seated are G. C. Druce, Mrs. Tansley, unidentified woman. Standing: unknown, J. Massart, C. Schroter, C. H. Ostenfield, Arthur Tansley, unknown, H. J. Cowles, L. A. Rubel, O. Drude, unknown, Mrs. Cowles, unknown, unknown, W. G. Smith, unknown, C. A. M. Lindeman. Individuals identified by John Sheail (Sheail, 1987). Photograph courtesy of the Clements Collection, American Heritage Center, University of Wyoming

having been part of a circle of younger men around the general. Phillips was not, however, personally acquainted with Tansley. He did know that Tansley and Clements were professional friends who had become acquainted through the International Phytogeographical Excursions of 1911 and 1913 that had taken place in England and North America (figs. 2.2 and 2.3) and that they and their wives corresponded regularly. Although Tansley accepted some of Clements's interpretations of vegetation patterns and introduced and used them in his work on British vegetation, he distanced himself from Clements's more extreme interpretations of the community as a complex organism and as a developing organism through ecological succession. Tansley made an attempt to accommodate Clements, but he was philosophically opposed to extreme speculation and arid intellectual taxonomy. Phillips would have done well to leave Tansley out of his pantheon of authorities.

2.3 Participants of the International Phytogeographical Excursion, in Yosemite, California, 1913. Seated, left to right: J. Massart, Arthur Tansley, Fisher, L. A. Rubel, behind, Skottesberg. Standing: unknown, Mrs. Tansley, Mrs. Brockman-Jerosch, H. J Cowles, C. Schroter, Engler, Edith Clements, with Frederic Clements (with hat) behind them. Others unknown. Photograph courtesy of the Clements Collection, American Heritage Center, University of Wyoming

Clearly, Tansley was offended by Phillips's articles. He was motivated to write *Use and Abuse of Vegetation Concepts and Terms* for the ecological community partly for scientific reasons concerned with the nature of the evidence and partly from the need to defend ecology from a too extreme philosophizing and to maintain its connection to mechanistic, reductionistic science and therefore its reputation within biology. In fulfilling this latter objective, Tansley's senses were sharpened by his personal difficulty in establishing ecology as a respectable discipline. Tansley was also put off by Phillips's mode of presentation, which seemed analytical but was actually quite arbitrary. We can understand how his statements caused Tansley to imply that they represented a closed system of religious or philosophical dogma. Tansley commented in an uncharacteristically harsh manner: "Clements appears as the major prophet and Phillips as the chief apostle, with the true apostolic fervor in abundant measure."[7] Even though "the *odium theologicum* is entirely absent," he wrote, "indeed the views of opponents are set out most fully and fairly, and the heresiarchs, and even the infidels, are treated with perfect courtesy" (Tansley, 1926, 677). Thus, Tansley's impatience with Phillips's articles probably stemmed as much from their underlying philosophical structure, which seemed to shape Phillips's presentation, as from their scientific content.

Although Tansley's article was concerned mainly with the terms and concepts reviewed by Phillips—succession, development and the quasi-organism, climaxes, the complex organism, and biotic factors—the new idea he offered was the ecosystem. In his presentation of the ecosystem concept, Tansley emphasized its physical character and its relation to physical systems in general. He made a decided effort to avoid biological metaphors and analogies. Although he stressed that ecosystems involved the interaction of the biota (he used the term *biome*) and the environment, he also placed the ecosystem within a larger ecological context. This larger context was the *formation,* usually defined as the regional vegetation adjusted to broad soil and climate patterns. Tansley used the formation to represent the climate, which he said contributed parts or components to the system together with the soil and organisms. Finally, the ecosystem concept he presented was part of several discussions of physical and ecological equilibrium.

Probably Tansley's use of equilibrium would be most problematic to today's ecologist. His concept of dynamic equilibrium has several elements. First, systems closer to equilibrium are most likely to survive. Second, equilibrium develops slowly as systems become more highly integrated and adjusted. The climax represents the nearest approach to a "perfect dynamic equilibrium" possible to obtain under the given conditions. Third, the equilibrium is never perfect; its perfection is measured by its stability. Compared to chemical systems, ecosystems are not stable because of their unstable components of soil, climate, and organisms and also because they are vulnerable to invasion by components of other systems. Finally, while it is possible that there is continual change in the system components, Tansley counsels us to split up the process of change and focus on the phases that are dependent upon the processes involved. Thus, Tansley's ecosystem concept is a physical concept, based on the concept of equilibrium and emphasizing the interaction of physical-chemical and biological components.

Tansley's offering to the Cowles festschrift was designed to be an analysis of ecological language, leading to a new alternative that would maintain ecology's links with the modern physical sciences and also create further bridges between alternative ecological approaches. His strategy was to address a variety of conceptual problems concerned with the nature of ecological communities and the consequences of speculative thinking in ecology.

THE ECOLOGICAL CONTEXT OF THE TANSLEY TREATISE

For at least a hundred years before the Phillips-Tansley exchange took place, field ecologists had been observing the patterns of organisms in nature and trying to understand how these patterns were formed and maintained by the interaction of the biota with the physical environment. The history of the development of ecological community concepts is one contextual element of the exchange. Tansley had been active in this type of research. He had organized the survey of British vegetation in the first decade of this century and had written *Types of British Vegetation* in 1911. In his response to Phillips and in his analysis of vegetation terms he was speaking from deep experience.

Depending upon the focus and scale of observation, the ecologist examining the patterns of vegetation might see an individual organism, a population of individuals, or aggregations of organisms in the form of forests, fields, and coastal banks. It was commonplace for nonscientists to recognize that nature was organized into forests, meadows, bogs, lakes, and so on. The early ecologists who focused on these commonplace units of nature usually presented their findings as lists of the species made up the biota of the unit, using the methods of Linnaean taxonomy. Ecological observation led to establishing lists of the species encountered on the survey. Ecologists then speculated about why particular species were present or absent, or why they were represented abundantly or rarely.

There are many examples of this early type of vegetation analysis, but the delightful report of Anton Kerner von Marilaun (1831–98), professor of botany at the University of Vienna, of his mid-century travels through the Danube Basin (1863) conveys especially well the character of these writings. In the following extract Kerner is thinking about the reasons why organisms are aggregated into communities:

> Wherever the reign of nature is not disturbed by human interference the different plant-species join together in communities, each of which has a characteristic form, and constitutes a feature in the landscape of which it is a part. These communities are distributed and grouped together in a great variety of ways, and, like the lines on a man's face, they give a particular impress to the land where they grow. The species of which a community is composed may belong to the most widely different natural groups of plants. The reason for their living together does not lie in their being of common origin, but in the nature of the habitat. They are forced into companionship not by any affinity to one another but by the fact that their vital necessities are the same. It may perhaps be true that

amongst the many thousands of plants inhabiting the earth no two are to be found which are completely alike in their requirements in respect of the intensity and duration of solar illumination, the concurrence of a particular duration of daylight with a certain amount of heat, the composition and quantity of the nutrient salts available at the places where the plants live, the amount of moisture in the air and in the ground, or, lastly, the character of the rainfall. This does not, however, exclude the possibility that in particular places similar demands may be met, and that different species with similar needs may flourish undisturbed side by side as men live together in one house or in one town, and, although their customs and their needs may not be exactly the same, yet form a society which is permanent and thrives, and wherein each member feels at home, because it rests upon common usages and is adapted to the local conditions. Nor is it impossible that each one may derive an advantage from the common life, that the associated individuals may support one another in the conduct of their lives, and that they may even be dependent upon one another. (Kerner, 1897, 885)

Kerner's interpretation of his observations introduce several important features of the genre. First, Kerner has focused on the plants in the passage quoted. This focus is evidence that ecologists were trained as plant or animal biologists and therefore tended to study plants or animals exclusively. Although partly owing to their training in separate disciplines in biology, this forced separation was due to the difficulty of mastering the taxonomy of several groups of organisms. As Phillips emphasized, before nature could be studied as a system it was necessary that all the parts be included.

Second, Kerner interpreted these groupings of plants to be communities, thinking analogically from human communities. In the writings of many authors at this time nature was being interpreted through the metaphor of a human community. Kerner was using the community metaphor in its positive, almost idyllic form, ignoring the reality of the human community in which he lived. At this particular time Europe was undergoing rapid social change as a result of industrialization, urbanization, and intense nationalism. In reaction, there was an emphasis on traditional, idealistic community forms and values, especially in the German romantic tradition, and these forms and values probably underlay Kerner's use of the metaphor. Social Darwinism and the human state interpreted as continual warfare was to come later.

Third, Kerner advanced two explanations for community formation. On one side, the community may be formed by a joint interchange and interaction between coexisting individuals. The community grows out of cooperation and competition between individual organisms. On the other side, they may be forced into companionship not by any affinity but by their joint needs, which are met in that particular habitat.

At the time of Kerner, ecological investigation was almost entirely observational. The ecologist walked or rode across a community and observed the presence, absence, and abundance of various organisms and interpreted those patterns. By the end of the century, however, ecologists were beginning to use quantitative methods in their investigations. For example, at the University of Nebraska, Roscoe Pound, later to be a famous jurist, and Clements applied a standard method of data collection derived from methods used by the German plant ecologist Oscar Drude and reported in his book *Deutschlands Pflanzengeographie*. Pond and Clements used square quadrats, five meters on a side, to define a limited plot of ground and recorded the frequency and abundance of individuals in those areas. By taking data from many quadrats within a community, they were able to observe variations in plant composition that were not visible to the naked eye and create quantitatively defined ecological units.

The ecologist using quantitative methods might conclude that the ecological units defined so precisely were concrete and fixed. Pound and Clements in their report on the phytogeography of Nebraska expressed this opinion:

> The vegetation of the earth's surface is arranged into groups of definite constitution and of more or less definite limits. Such a group is a plant formation. It is necessary to distinguish very carefully between formations and minor groups, facies, and mere patches. A formation is invariably a plant-complex, except in its incipience or decadence. It has to do primarily with the species which compose it, though these are represented in it necessarily by individuals, while a facies or a patch derives its character solely from the individuals of its species. . . . The plant formation determines not only the constitution of the floral covering, but is also a more or less interpretable expression of those biological forces of which it is a resultant. It is a biological community in which each factor has more or less interrelation with every other factor, a relation determined not merely, nor necessarily, by the fact of association but also as a result of biological forces induced by physiographical and meteorological phenomena. (Pound and Clements, 1897, 313–14)

In Pound and Clements's explanation of the patterns of plant distribution, they mention the role of both biotic interaction and the physical environment. Later, these factors became transformed in the Clementsian theory of development of the complex organism. As Phillips stated, the climate set the type of the climax vegetation, while development was entirely a process of biotic interaction. Obviously, there could be a variety of interpretations of the causal processes forming communities. The ecologists conclusions were strongly influenced by the information they obtained from their quadrats and surveys.

The Evidence of Field Studies

In ecological field studies, whether one focuses on plants, animals, or both, the ecologist is faced with interpreting the presence and abundance of hundreds (even thousands) of species consisting of possibly tens of thousands of individuals. Since each species has evolved to fit certain environmental conditions, the of variability in biotic presence and abundance is enormous. Our capacity to understand such patterns depends partly upon our mathematical tools and the availability of instruments to process quantitative data. At the time of Clements, Phillips, and Tansley, these tools were relatively crude. It was possible to see patterns relatively clearly in certain circumstances, especially where the environment was so harsh that it limited the species that could live there. Yet in more temperate or tropical conditions the patterns tended to be confused by gradients of distribution across varying environments.

Environmental conditions varied across space, and individuals of species present in the species pool of the region could disperse, germinate, and grow on those gradients. Individuals would grow and mature in that part of the gradient where the conditions were adequate to meet their requirements, unless other organisms occupied the space. A species would likely be most abundant under the conditions most closely fitting its requirements. Even so, the distributions of species were seldom precise or predictable because both the environmental factors and the biota were discontinuously distributed and were changing dynamically in space and time.

Theoretically, a thorough knowledge of the physiological and behavioral requirements of species would permit us to predict where they would occur. Shelford, author of the influential book *Animal Communities in Temperate America as illustrated in the Chicago Region* used this general approach in his studies of animal communities with considerable success. A consequence of environmental gradients, however, was that communities were seldom, if ever, precisely defined patches made up of constantly occurring species in constant ratios of abundance. Rather, if one was inclined to accept the existence of communities in the first place, each community seemed to be an single example of a type.

A further spatial problem with community analysis, which is the obverse of the problem of deciding who are the residents of a community, was deciding where community boundaries occurred. The distribution of species in continua across environmental gradients meant that the boundaries of communities were often fuzzy and imprecise. Tansley, in *The British Islands and Their Vegetation,* recognized the problem:

> Much vegetation too, particularly in countries subject to varied human activity, is difficult or impossible to separate into distinct, well-charac-terized communities—it often presents all grades of mixtures of the ele-ments of several communities in which dominance and layering are con-fused, obscure, or totally absent. . . . As a general principle, the longer the vegetation is let alone and left to develop naturally, the more it tends to form well-defined communities, and the more these develop relatively constant and well-defined "structures" in relatively stable equilibrium with their conditions of life. (Tansley, 1939, 215)

Pound and Clements in *The Phytogeography of Nebraska* developed a practi-cal approach to the boundary problem, based on their observation of plant distribution in the field. They stated in the section in their book on plant formations:

> Defined accurately, a formation is a piece of the floral covering, the ex-tent of which is determined by a characteristic correlation or association of vegetable organisms, i.e. it is a stretch of land the limits of which are biological, and not physiographical. It can rarely have definite limits, therefore, but must be bounded on every side by a more or less extensive belt in which the features of two adjacent formations are confused. As in the case of species, it often becomes necessary to establish arbitrary lim-its, within which the preponderance of characteristics must be adopted as the mark of delimitation. (Pound and Clements, 1897, 315)

Pound and Clements were dealing with vegetation in the western United States prairies before that area was greatly altered by agricultural development. They were able to see patterns of gradual change from one community to another.[8] Further, their quadrat data described continuous change in the fre-quency and abundance of species. Under such circumstances, boundaries are indistinct except where a discrete physical factor is present. In contrast to the unplowed American prairie, the European situation was quite different. There, ecologists were faced with a patchy landscape, long under human control, which was frequently well mapped for military purposes. Owners of land could describe with considerable accuracy the locations of forests, meadows, ponds, and similar communities. Indeed, the ecological historian Oliver Rackham has shown in England that some forest stands persisted from Roman times to the recent period (Rackham, 1980). Under conditions of long-term persistence, the boundary problem may not be difficult to resolve and attention can focus on other matters. Josias Braun-Blanquet (1884–1980), the phytosociologist, for example, recognized that boundaries may be wavy or indistinct (Braun-Blanquet, 1932, 77), but in his method, he placed the quadrats used to describe the stand at the center of a representative section of the community.

In community ecology the ecologist is also dealing with two different kinds of temporal variables. First are environmental features that have temporal dynamics ranging from the day-to-day changes in the moisture conditions of surface soil to geological uplift and erosion that may operate on time frames of thousands or millions of years. Second are the temporal responses of the living organisms that occupy the environment. These responses may range from ancient trees that live hundreds of years to organisms with life cycles that last hours or days. The intersection of these time patterns also results in a great diversity of potential forms and patterns.

As a consequence, the ecologist observes that the communities change over time as well as over space. Where the process is geological in scale, as after glaciation, the communities gradually shift as species invade and disappear, depending upon their capacity to move and interact and on the changing environmental conditions of the site. Where the land has been disturbed by a volcanic irruption, fire, storm, or human disturbance, the site is unoccupied at the start. Organisms occupy the site in waves of invasion and settlement, and as a consequence there appears to be a transition of communities replacing each other.

Development of the Complex Organism Concept

These brief comments on ecological observation of communities suggest that nature may be organized but that close analysis reveals the tremendous complexity of species presence, variability from place to place, and a great deal of change locally in abundance and presence. Where the environment is especially harsh, the patterns may be more orderly, visible, and repeatable. In temperate and tropical environments, however, the environmental conditions are less restricting and biotic interactions make it more difficult to predict and explain community structure.

Several paths through this jungle of complexity, variability, and multiple causation were invented by ecologists. First, a hierarchical approach, copied in part from the taxonomists, was imposed on natural communities. A regional pattern was recognized, representing the most common life form of vegetation within a region—for example, a forest formation under humid climates, or a grassland or desert under a more arid climate. Second, within a region, there is a mosaic of communities of different life forms. In a forested region, besides forests, one could encounter meadows and bogs, which are caused by local environments under the control of soil, topography, and water level. Finally, if you focus on a single community patch, that patch usually has a particular species composition and abundance that differs from that of the next patch. These local patterns frequently are influenced by the interaction of the biota and the chance occurrence of species.

These patterns create a top-down and bottom-up problem for the interpreter. If one approaches nature from the top, as I have explained in the above paragraph, then one analyzes the pattern by subdividing it and looking for an explanation for each division. The analyst asks, what makes these communities different? How different must the communities be before considering whether there are two or more types of communities in the sample? In the top-down approach the analyst seeks criteria for dividing the whole into its component parts.

Alternatively, if the scientist takes the bottom-up approach, then he or she begins with a collection of individual communities and asks questions about what they have in common. After the criteria for combination are devised, then the ecologist can organize the communities into patterns. The higher units are abstractions that are characterized by the common properties of all the samples but that mimic no individual sample. In this way the ecologist might create an ecological taxonomy, in the same way a taxonomist defines units such as genera, families, and orders that represent hypothesized phylogenetic relationships among the species. Scientists using the top-down and the bottom-up strategies see the world differently and tend to argue strongly about their interpretations.

Ecological science at the time of Phillips's and Tansley's publications was engaged in one these arguments. There were those who saw in the formation, the higher-order abstraction that represented the regional vegetation, as a reality that overlay the finer patterns of the community patches. There were those who focused on the actual stands of vegetation and tried to organize these stands into patterns. This latter group was subdivided into those placing greatest reliance on the presence or absence of species and those focusing on the environmental factors that selected for species. Practically all ecologists of the period were touched by these arguments.

In the United States the interpretation of the vegetation patterns was dominated by the thought of Clements. Clements was raised and educated in the frontier state of Nebraska, and he experienced the great American prairie as a botanical abstraction—the prairie unplowed without bison or native Americans. Clements traveled extensively across the United States as a researcher of the Carnegie Institution, chauffeured by his wife, Edith, who was a trained botanist in her own right. Based on his experience and observations, Clements created a theory of vegetation that was reported in a set of volumes published from 1905 to 1939. The most important of these for this story was *Plant Succession,* published in 1916.

Clements viewed a region as having a characteristic vegetation, called the climax, which was caused by the selection of the regional climate for particular life forms of plants. Of course, Clements observed a variety of communities within a region. He interpreted these different communities through a theory of change or, as he called it, development. The technical term used for this process was *ecological succession.* In Clements's successional theory, all of the stands of vegetation were on trajectories of change converging on the climax type. In some communities change was rapid, but in others change was so slow that the communities might warrant being given a modified climax term, such as a *disclimax* or an *edaphic climax.* Clements invented a complex terminology to fit his theory to what he observed.

Clements went further than a descriptive theory by drawing an analogy between the climax community and an organism. He called the climax a complex organism to distinguish it from the well-recognized individual organism, commonly used in biology. He then asserted that the complex organism went through a life cycle of birth, growth, and development.

Clements had created an awe-inspiring concept of nature. His invention was deterministic, all-inclusive, and internally logical. Even though many did not agree with him, he was an effective advocate for his ideas. Further, as a scientist involved in active field work, he assimilated some critical comments and considerably modified his theory over the years.

In most community studies and especially in the Clementsian theory of succession, the focus was on the biota and the biotic interactions and processes thought to control community dynamics. The environment was considered to be a secondary factor; frequently it was called a stage on which the biota acted a drama. It was implied that living organisms responded to environmental conditions through their physiology and behavior, and in doing so they also influenced and changed the environmental conditions through their "reaction" on and change of physical-chemical factors. Since ecology was a biological subject, it was unlikely that ecologists would be able or interested in examining the environment deeply from a physical-chemical viewpoint. For this reason, there was almost no attempt made to relate the observations of chemists, especially geochemists such as Vladimir Vernadsky and other Russians, to community ecology. Thus, Tansley's emphasis on the interaction of the biota and the environment in the ecosystem was an important conceptual advance and opened the door for the wider use of energy theory and matter cycling in ecology.

SPECULATION ABOUT ECOLOGICAL OBJECTS

It is clear that the interpretation of the patterns of natural communities included an element of abstraction that went beyond the evidence of field observation. The employment of the metaphors of the human community and the complex organism to describe ecological objects and patterns has been typical of the subject. To understand the development of ecological science it is necessary to understand its language and the philosophical concepts that form a deeper context of the subject and its practitioners. For example, Ronald Tobey (1981), in his study of Clements, commented that the use of the organism metaphor for human society by Herbert Spencer and Lester F. Ward in their widely read books probably influenced Clements to use the organism metaphor for his climax community.

Where there were little or no data on mechanisms or experimental experience, ecologists have turned to other sciences and philosophy for allied concepts in interpreting their observations. In assessing the phenomenon it is important to remember that isolated disciplines and specialists were less prevalent in times past. Ecologists shared with other scholars a background in education, which included the study of classical civilizations, languages and literature, and science and mathematics. Thus, as research advanced in new directions, it was easy to dip into a common pool for metaphors that drew connections between what was known and what was new. This approach has tended to be a valuable one for ecological researchers.

John Phillips sought support for the concept of the complex organism from philosophy. He turned to a concept created by Jan Smuts, who had presented his philosophical thought in a 1926 book, *Holism and Evolution,* wherein he called the synthesis of matter, life, and mind—based upon science—*holism*.[9] Smuts noted that philosophers and scientists tended to treat matter, life, and mind as separate phenomena, arguing that they "will appear as a more or less connected progressive series of the same great Process. And this Process will be shown to underlie and explain the characters of all three, and to give to Evolution, both inorganic and organic, a fundamental continuity which it does not seem to possess according to current scientific and philosophical ideas" (Smuts, 1926, 21).

He identified unified structures, which he called wholes, that included physical bodies, chemical compounds, organisms, minds, and personalities. The operative factor which creates wholes is a process of creative synthesis, in which the whole is the synthesis of the parts. For example:

> Taking a plant or animal as a type of a whole, we notice the fundamental holistic characters as a unity of parts which is so close and intense as to be

more than the sum of its parts; which not only gives a particular conformation or structure to the parts but so relates and determines them in their synthesis that their functions are altered; the synthesis affects and determines the parts, so that they function toward the "whole"; and the whole and the parts therefore reciprocally influence and determine each other, and appear more or less to merge their individual characters. (Ibid., 86)

Smuts stressed that the whole is not a simple object, but is complex, consisting of many interacting parts. The parts themselves also may be wholes, as the cells in the body of an organism are wholes. The whole is not a mechanical system, which he characterized as one lacking inward tendencies and where all action is external. All action is through the machine acting on external objects or the action of external objects on the machine. Wholes, in contrast, have inner tendencies that produce more than a machine. Finally, Smuts stressed that wholes are not additional to parts; wholes are the parts in a definite structural arrangement with reciprocal activity and function. To support this idea he used the familiar story of hydrogen and oxygen as chemical compounds with unique properties, which, when combined to form water, have new unique properties as water.

Smuts argued that the concept of the whole transformed the concept of causality: "When an external cause acts on a whole, the resultant effect is not merely traceable to the cause, but has become transformed in the process. The whole seems to absorb and metabolize the external stimulus and to assimilate it into its own activity; and the resultant response is no longer the passive effect of the stimulus or cause, but appears as the activity of the whole" (ibid., 119). Thus, the whole appears as the cause of the external response.

Smuts did not treat the ecological community as a whole, although he did consider human society, families, and nations as wholes. He considered nature to be made up of wholes but was careful to distinguish between nature as an organism, which he denied, and nature as organic through the intensification of the entire field. In Smuts's words, "Nature is holistic without being a real whole" (ibid., 340). He attributed this holistic force of nature not only to humans but to all organisms: "The new science of Ecology is simply a recognition of the fact that all organisms feel the force and moulding effect of their environment as a whole. There is much more in Ecology than merely the striking down of the unfit by way of Natural Selection" (ibid., 340).

The extension of Smuts's holism to the ecological community appears to be Phillips's creation. If Phillips had not used Smuts's thought in the way he did, ecologists probably would have little interest in Jan Christian Smuts or his philosophy.

Conceptual Differences

Phillips's use of Smuts's concept of holism as support for his ecological speculation introduces another problem area that Tansley was addressing with his ecosystem concept. This problem area is complex, and we need to sort out several arguments in order to understand it and use it in this analysis. There are two contrasting arguments that underpin ecological speculation. First is the contrast between materialism and idealism, which was especially active in science and philosophy in the nineteenth century. The materialists argued that all phenomena are material and therefore, potentially at least, able to be understood by applying the methods of science. Ernst Haeckel was a famous representative of this viewpoint. Haeckel was well known as an advocate of Darwin's evolutionary theory and was a vigorous opponent of all spiritual theories and beliefs. The idealist took the opposite view, arguing that there were phenomena that were immaterial and could not be penetrated by science. Religious idealism, of course, posited a God who was above materialism, but scientific idealists also tilted against the materialists. Idealism was an apparently reasonable position to take when little was known about a phenomenon or when a phenomenon was beyond analysis by the scientific methods available at the time.

At the end of the nineteenth century, when Tansley was being educated, there was a well-known debate in biology about an idealistic concept called *vitalism*. The problem involved the nature of life. Although one could analyze a living organism into parts, such as tissues and cells, there was no way to reconstruct the parts into a whole organism and have life. Life seemed to be something immaterial; life was said to represent a vital essence.[10]

Vitalistic concepts were frequently used in biology to explain phenomena that appeared to be unexplainable in materialist terms. One after another, however, these phenomena were explained by conventional research founded on materialist principles, and the vitalist argument was gradually discredited, being held mainly by those defending a religious interpretation of biology. Tansley would very likely have been exposed to these arguments as a student and probably was conditioned to be suspicious of idealist arguments such as holism or the complex organism.

In the ecological story we are considering here the materialist-idealist contrast would be between the description of vegetation on a landscape—such as described by Kerner or Pound and Clements—made up of communities with the observed species compositions, contrasted to Clements's interpretation of vegetation as an organism that grows, matures, and dies. The concepts of the complex organism or the superorganism are idealist concepts that are not researchable using ecological methods of analysis. Certainly, this was one reason for Tansley to argue against the concepts.

The second argument is between reductionism and holism. Reductionism claims that we can understand the nature of a phenomenon by reducing it to its parts. Analysis of these parts reveals the mechanism of the phenomenon. An engineer will take a reductionist approach when repairing a motor. In mechanical reductionism we are dealing with objects made by humans, for which a plan exists. The problem is to restore the parts to their proper order and link them together to fit the plan. In ecology we are studying phenomena not of our making and for which there is no plan. Research has to develop both the plan and the mechanism.

Holism takes another approach. Smuts's form of holism has been described, but holism can be described in a less idealistic form. Generally, the materialist holist is concerned with how parts are organized to create wholes; that is, the holist is concerned about the rules used to assemble parts into functional wholes. Usually, the scientific holist does not deny the value of reductionism, agreeing that it is necessary to understand the parts and how they act, but adds that it is also essential to understand the rules that are used to assemble the parts to make an object.

The argument between reductionism and holism is an interesting one in ecology because it is unbalanced. First, reductionism seems endless. John Harper introduced his E. P. Odum lecture at the University of Georgia in the mid-1980s by saying that in his research he was digging a hole. Deeper questions arose sequentially as Harper moved from the study of the vegetation of a Welch sheep pasture to the molecular genetics of clover. Because of this endless quest, the reductionist is impatient with the holist—the interesting questions move the reductionist away from synthesis and from the starting point of a particular research. On the one hand, as reductionism proceeds, the reductionist has less and less in common with the ecologist, who is concerned with the broader issues of organization and may even begin to question the value of ecological work in general. On the other hand, the holist understands that the results of reductionist research are always relevant at some level in understanding a phenomenon. Thus, the holist is tolerant of the reductionist agenda and even supportive of it, if the competition is not too keen.

Another problem can emerge in this contrast, which we see in the Phillips's articles. The reductionist continues using conventional methods to go deeper into the analysis. There is no need for the reductionist to use philosophical concepts to explain the findings of research. Philosophy is relevant to the location of the hole, not to the digging of it. In contrast, the holist has made little progress in creating assembly rules. I have already noted the problems of complexity, diversity, change through the environment, and evolution in the natural community—all of which make the ecological system unstable in Tansley's terms. The holist, then, may be tempted to reach for other support for a theory of organization. Usually, the support is a metaphor and the argument is developed analogically. An example of this form of thinking is that if the world is a heat engine, then x and y follow. If you do not accept the metaphor of the world heat engine, you will not accept the logic that follows. Analogical thinking is valuable to establish new hypotheses to follow in research in an area where little is known. It is less valuable where the research plan is clear.

The complexity of the problem in the philosophy of science suffices to show the underlying problem in the Phillips's articles. Both Phillips and Tansley were materialists. Phillips was a holistic materialist, using Smuts philosophy of holism to support the concept of the complex organism. Tansley, aware of the split between reductionist and holistic materialism, which divided ecology into two unreconcilable parts, sought a common ground. His ecosystem concept was offered as a bridge. In developing the ecosystem concept he avoided biological and organismic theories altogether. They were a trap that led to emotional debates between biologists and ecologists. Rather, he presented a physical theory that was founded on the concept of equilibrium. Although Tansley does not say so directly, equilibrium and stability provide the foundation for assembly rules. The more stable the system, the more likely it is to persist; systems tend to move toward equilibrium, and so on.

Whitehead's Science and the Modern World

In coming to this position Tansley is not unique, although his term, *ecosystem*, is unique. Many other scientists and philosophers were trying to find links and bridges that would prevent intellectual life from shattering into parts that could not communicate with each other. One thinker that was very close to Tansley in his ideas and who was referred to by Tansley and rejected as an "organicist" was Alfred North Whitehead (1861–1947). Eugene Hargrove pointed out in his book *Foundations of Environmental Ethics* that Whitehead's holistic concepts also were very near those of Aldo Leopold (1887–1948), the famed wildlife biologist and conservationist from Wisconsin who at that time was developing a powerful statement of practical ethics. In Leopold's *Sand County Almanac*, he stressed that "a thing is right when it tends to preserve the integrity, stability, and beauty of the biotic community. It is wrong when it tends otherwise" (Leopold, 1949, 224–25).

Whitehead much more than Smuts provided a philosophical ground for the holistic ecological concepts Leopold and others advanced. In 1925, Whitehead presented the Lowell Lectures at Harvard, which were published together with some additional essays as *Science and the Modern World*. In this book

Whitehead examined the history of Western thought, especially the interaction of science and philosophy. He contrasted the pattern of medieval thought, which assumed a world of order created by God and understandable to the rational mind, with the modern concept of a particulate, materialistic world in which order, if it exists, is relative and momentary. In these systems every detail and event was supervised and ordered, and the search into nature vindicated a faith in rational organization. Further, every occurrence could be correlated with its antecedents in a definite manner, exemplifying general principles. The medieval world's interest in nature began in the arts, for example, in the curling tendrils of vegetation on cathedral columns, and possibly more powerfully through the monastic experiments in agriculture and forestry, which were among the first steps toward a modern scientific view. In the Renaissance the focus shifted to the immediate, simple fact of observation and experience and the belief that individual humans and higher animals were self-determining organisms. This shift created a profound revolution in thought and made modern science possible.

Like Tansley, Whitehead recognized the contradiction in science, which stressed the primacy of fact and individual interpretation of fact, yet had the grander purpose of discovering patterns or order in the natural world. According to Whitehead, this contrast between a deterministic and relativistic interpretation of particular events, as well as generalizations leading to broad conclusions based on rational thought, was a general problem for science and culture. While the contradiction might be avoided within the physical and chemical sciences because the wholes recognized by these scientists remained physical, the problem was central to the biological and social sciences. In ecology the contrast was illustrated by the determinist theory of Clementsian succession, cited by Phillips, as contrasted to the relativistic approach advocated by Henry Gleason (1882–1975) of the University of Illinois. Gleason (1926a, 1926b, 1939) stressed the role of individual plants to invade and grow on the environmental gradients and declared the plant association was a "mere coincidence."

Whitehead pointed out that these points of view resulted in three positions: dualism, in which both individualism and holism coexist, or monism, which in turn places individualism within holism, or holism within individualism. Whitehead accepted none of these. Rather, he proposed a bridge position designed to solve the contradiction: "The doctrine which I am maintaining is that the whole concept of materialism only applies to very abstract entities, the products of logical discernment. The concrete enduring entities are organisms, so that the plan of the *whole* influences the very characters of the various subordinate organisms which enter into it. In the case of an animal, the mental states enter into the plan of the total organism and thus modify the plans of the successive subordinate organisms until the ultimate smallest organisms, such as electrons, are reached" (Whitehead, 1944, 115). For Whitehead, biology was the study of larger organisms, and physics was the study of smaller organisms. The character of these organisms, Whitehead postulated, was the "event," the ultimate unit of natural occurrence. Tansley had used a similar metaphor in formulating the ecosystem concept, derived apparently from a popular book on science, *The Universe of Science,* by H. Levy. Two patterns characterize an event: "Namely, the pattern of aspects of other events which it grasps into its own unity, and the pattern of its aspects which other events severally grasp into their unities" (ibid., 174).

The event had a temporal aspect, an endurance, a retention of form or value over time. It repeated the shape exhibited by the flux of its parts, so the entity had a life history represented by the dynamics of its parts. Additionally, the life history of the individual entity was part of a larger, deeper, more complete pattern and may be dominated by aspects of this larger pattern. Whitehead called this the *theory of organic mechanism.* Thus, he concluded that there were two aspects involved in the development of nature:

> On one side, there is a given environment with organisms adapting them-
> selves to it. . . . From this point of view, there is a given amount of mate-
> rial, and only a limited number of organisms can take advantage of it. The
> givenness of the environment dominates everything. . . . The other side of
> the evolutionary machinery, the neglected side, is expressed by the word
> *creativeness.* The organisms can create their own environment. For this
> purpose the single organism is almost helpless. The adequate forces re-
> quire societies of cooperating organisms. (Ibid., 163)

In these writing Whitehead used the theory of biological evolution to provide a motive force to create dynamic interaction. Tansley used the physical concept of equilibrium to serve this purpose.

The System Concept

It remains to explore Tansley's use of the concept of system in his use of the term *ecosystem.* Tansley used the systems idea frequently in his writing. It served to organize his concern about the organization of nature. For example, as F. F. Blackman's and Tansley's review of Clements's *Research Methods in Ecology* explained:

> The book is divided into four parts or "chapters." The first, under the
> heading "The Foundations of Ecology," discusses "The Need of a Sys-
> tem," and contains the ideas upon which we have already commented.

We think Dr. Clements easily establishes his case for a "system" and his views of "The Essentials of a System," based on the absolutely fundamental importance of the habitat, its effect on the plant and the reaction of a plant upon it, may be said to be almost self evident. . . . The ecological investigator in the midst of vegetation finds himself in the presence of a state of equilibrium between the organized and the unorganized, between "The Habitat" of the one part and "The Formation" of the other part. (Blackman and Tansley, 1904, 203, 232)

In 1911, in *Types of British Vegetation*, Tansley commented;

It may be said that we ought not to occupy ourselves with synecology till we have a complete or an approximately complete knowledge of autecology, but this is a mistaken notion. It might as reasonably be contended that we ought not to study the phenomena presented by the nations and races of men before we know all about the physiology and psychology of the individual man. As a matter of fact the study of synecology is considerably in advance of autecology (which is indeed still in a very backward state of development) and the progress made has amply justified the attention devoted to the wider though less fundamental branch of the subject. (Tansley, 1911)

In 1926, in *Aims and Methods in the Study of Vegetation*, Tansley and T. F. Chipp (p. 141) made the point:

How are we to draw the line between those which act as *members* of the community and those which are to be considered external—whether "hostile" or "friendly"—to it? Some ecologists have tried to get over the difficulty by considering *all* the organisms, animals and plants together, living in one place and mutually acting upon one another, as members of the community, as a biotic unit. There is a good deal to be said for this conception from a philosophical point of view, for it is really the whole of the living organisms together, *plus* the inorganic factors working upon them, which make up, in a climax community, a "system" in more or less a stable equilibrium. But such a "system" considered fundamentally, that is, physically, must include the "inorganic" factors of the habitat and these obviously cannot be considered as "members" of the community, and if we take the inorganic factors as external, why not biotic factors such as grazing animals?

The concept of a system is ancient and had wide use in science. In conventional terminology system meant a complex in which the parts interacted to produce the behavior of the whole. In the period between the two world wars, however, system science emerged as a separate subject. Ludwig von Bertalanffy (1901–72), a major figure in German theoretical biology (Davidson, 1983), describes this period of development of the system concept (Bertalanffy, 1952): "The future historian of our time will note as a remarkable phenomenon that, since the time of the first World War, similar conceptions about nature, life and society arose independently, not only in different sciences, but also in different countries. Everywhere we find the same leading motifs; the concepts of organization showing new characteristics and laws at each level, those of the dynamic nature of, and the antitheses within reality. (Bertalanffy, 1952)

Bertalanffy traces the development of a philosophy of systems dynamics from Heraclitus, the sixth-century B.C. Greek philosopher noted for his ideas on universal flux, paradox, and the unity of all things;[11] Cardinal Nicholas of Cusa in the Italian-German renaissance, who taught the infinity of the universe and the coincidence of opposites; and Goethe, the father of morphology, to the philosopher Nicolai Hartman, who in 1921 described a system whereby forces balance one another and lead to a stable configuration, which is organized in a hierarchical pattern. Bertalanffy's list of founders could be expanded to include other philosophers and scientists, depending upon the aspect of systems theory one wished to emphasize. For example, the American ecologist Charles Christopher Adams expressed almost identical thoughts to those of Bertalanffy but derived from Herbert Spencer in 1915 and 1918 publications.

Systems science was a technical application of holistic, materialistic philosophy. Paraphrasing the systems scientist Mario Bunge (1979), the typical scientist or engineer applies a particular science to the general problem, but the systems expert deemphasizes the physics, chemistry, biology, or sociology of his or her system, focusing instead on its structure and behavior and the possibility of duplicating this behavior with that of a system of a different kind. It was possible to develop conceptual models of system behavior in the between-the-world-wars period, but operative, manipulable models required mechanical computers, which were years in the future. Thus, systems science in this period had a focus—the discovery of the assembly rules through which the parts of systems could be organized into wholes—and an objective—to connect isolated systems into networks, but lacked a method.

Sir Arthur George Tansley proposed the concept of the ecosystem as a solution to several vexing conceptual problems in ecology, introducing the term in his contribution to a festschrift for Henry J. Cowles. Tansley was disturbed by a set of articles by the South African plant ecologist John Phillips that illustrated in Tansley's mind several of the problems he wished to counteract.

Tansley's ecosystem concept was a physical concept that stressed that both the physical-chemical environment and biotic organisms acted together to form an ecosystem, which was in turn formed part of a hierarchy of physical systems from the universe to the atom. The physical concept of equilibrium guided the organization and maintenance of ecosystems. The stability or persistence of the system involved its movement toward equilibrium. Tansley was quite modern in entertaining the possibility that ecosystems seldom, if ever, achieved stability.

The ecosystem concept emerged in a theoretical argument. It was not the result of a technical study and was not presented as the synthesis of field observations. Further, Tansley never used the ecosystem concept in his studies, although he did use the concept in later conceptual writings. Rather, the term emerged at a time receptive to such concepts. Systems science had begun its development, and while Tansley distanced himself from Alfred Lotka, the American physical biologist, and Bertalanffy's efforts to create a biological systems theory, his ecosystem concept fit the emerging pattern of systems science. Philosophers such as Whitehead were also formulating theory in terms similar to Tansley's ecosystem, but again Tansley did not avail of them to support his concept. Rather, he depended upon the connection of the ecosystem to the physical sciences, the most precise and mathematical sciences, and the concept of physical equilibrium to convince his audience.

Tansley did not clarify whether he considered an ecosystem to be an object of nature or something else. His event-focused physical orientation may have favored treating the ecosystem as an event in a physical field of dynamic process. Yet Tansley did not use the physical field concept in his presentation, nor did he refer to the flow of energy or matter cycles, both topics that would later become central elements of ecosystem studies. Because he was unclear on these points, ecologists tended to misuse the term *ecosystem* as a more modern expression for the community concept or Clementsian complex organism and thus maintained the confusion that Tansley was trying to overcome.

Tansley's strategy can be understood as a consequence of his frustrating experience to obtain recognition both for ecology as a serious science in Britain and for himself within British academic life as an ecologist with high standards for technical work and writing. He grew up in a time when the great discoveries in physics at the Cavendish Laboratory at Cambridge—close to the Botany School where Tansley taught—excited scientists everywhere. Thus, Tansley chose a familiar and acceptable authority on which to anchor his ecosystem concept.

CHAPTER 3

The Lake as a
Microcosm

I have suggested that Arthur Tansley formulated the ecosystem concept as a solution to a conceptual argument that divided plant community ecology into two opposing camps. One group emphasized the significance of the individual stand of vegetation and organized these stands into hierarchies of community organization. The other hypothesized that vegetation was a complex organism that developed, matured, and became senescent. Tansley was aware how such arguments could act against the reputation of a field of inquiry and was acutely conscious of the low esteem in which his fellow physiologists, morphologists, and geneticists held ecology. Part of Tansley's motivation in creating the ecosystem concept was a desire to find a bridge that would link these two points of view into one ecological approach.

I have also suggested that the language Tansley used for his concept of ecosystem was derived from the scientific and philosophical ideas current during the early twentieth century in England and the United States. Among these ideas was the concept of a system, as used widely in science and technology, and that of a physical equilibrium. Unlike the ecologist's community or complex organism, Tansley's ecosystem was composed of both the physical-chemical environment and the entire biota, not just the plants or animals. These concepts are related and together they represent a different aspect of the broader ideas that were derived from both the scientific interpretation of nature and holistic concepts of Western European philosophy. Although Tansley attached his ecosystem concept to the physical sciences, he carefully avoided going very far beyond scientific experience with his concept. Even so, he opened a door for those with a desire to find universals in nature to create a new ecological synthesis. As far as we can tell from the literature, he never employed the ecosystem concept in his own scientific work.

The tradition in which Tansley worked was terrestrial plant ecology. He initiated plant surveys of the British isles and wrote extensively on British vegetation. The ecosystem concept was presented as a solution to an argument about vegetation terms and concepts. Yet there were other ecological traditions that addressed these questions from other perspectives. One of the most important was fresh water ecology, or as François Alphonse Forel (1841–1912) named it in 1892, *limnology*.[1] Tansley was undoubtedly aware of the advances made in limnology since botanically oriented treatises on freshwater ecology were published frequently in the two technical journals he edited, *The New Phytologist* and the *Journal of Ecology*. Still, Tansley did not cite articles from this tradition in support of his ecosystem concept. Indeed, the aquatic ecologists were well ahead of terrestrial ecologists in developing an operative system concept.

One of the first of these aquatic system concepts was the idea of a lake as a microcosm, expressed by Stephen Alfred Forbes (1844–1930), a distinguished Midwest American biologist. Forbes founded the Illinois State Laboratory of Natural History (later the Illinois State Natural History Survey) in 1878, was appointed state entomologist in 1882 and head of the department of zoology and entomology at the University of Illinois in 1884. He was strongly influenced by Herbert Spencer,[2] especially in his employment of the concept of the balance of a system. Forbes described a lake as "an old and relatively primitive system, isolated from its surroundings. Within it matter circulates, and controls operate to produce an equilibrium comparable with that in a similar area of land. In this microcosm nothing can be fully understood until its relationship to the whole is clearly seen. . . . The lake appears as an organic system, a balance between building up and breaking down in which the struggle for existence and natural selection have produced an equilibrium, a 'community of interest,' between predator and prey." In this 1887 statement, made almost fifty years before Tansley formulated his concept of the ecosystem, Forbes anticipated several of the points Tansley was trying to make, and in one sense at least, went beyond Tansley. Forbes's vision of a lake as a isolated object, a system in which cycles of matter maintain an equilibrium between the forces of production and decomposition, has—more than one hundred years later—a contemporary cast.

The ideas of Forbes were not well known outside of the Midwest. Part of the reason for his obscurity was that he published in the *Peoria* (Illinois) *Science*

Association Bulletin (later republished in the 1920 *Bulletin of the Illinois Natural History Survey*). August Thienemann, who would likely have made use of Forbes's ideas, did not refer to him in his 1925 survey of European freshwaters, *Die Binnengewasser Mitteleuropas*. Another reason for the delay in implementing Forbes's microcosm concept was that the development of a functional approach required a strong descriptive base. It is necessary to know what organisms live within the system and to understand their linkages with the environment before they can be organized into a system. It was not until the period between the first and second world wars that enough information had been accumulated to take this step.

Forbes's 1887 image of the lake as a microcosm introduced another group of scientists who made a much greater contribution to the development of the ecosystem concept than did the terrestrial vegetation scientists. Limnologists were distributed across the northern latitudes coincident with the distribution of lakes on the landscape. The postglacial and mountain landscapes in North America, Europe, and Asia provided numerous opportunities for the formation of lakes, and as J. G. Needham and J. T. Lloyd showed in their 1916 textbook for North America, biological field stations developed on the shores of these habitats early in the twentieth century.

The scientific study of lakes began with Forel's three-volume study of Lake Geneva during the last decades of the nineteenth century. Work in limnology in general, and in lakes in particular, proceeded rapidly, so that by the early twentieth century there was a large amount of literature on the subject. In 1916 James G. Needham and J. T. Lloyd of Cornell University published an elementary textbook on fresh water biology, and in 1918 Henry B. Ward of the University of Illinois, and George C. Whipple of Harvard University, collaborated to produce their famous *Fresh Water Biology*. Both of these volumes were devoted mainly to the biology of individual taxa, although Needham and Lloyd discussed aquatic "societies" in their volume.[3]

The development of limnology advanced rapidly in the midwestern United States and northern Germany. In the United States the science was stimulated by the presence of lakes and the establishment of new universities. The University of Wisconsin was constructed on the shore of Lake Mendota and Cornell University was located on the Finger Lakes of New York. The University of Michigan was surrounded by many small glacial lakes. All of the new institutions became centers of limnological studies. The settlement of the midwestern states had occurred only a few years before, and the formation of universities provided opportunities for the organization of new subjects and approaches to education and research. For example, when Edward Asahel Birge (1851–1950), who organized the work in Wisconsin, began his professorship at the

University of Wisconsin in 1879, his was a one-man department of biology (Frey, 1963). He later became the first chairman of the department of zoology, the first director of the Wisconsin Geological and Natural History Survey, and eventually, president of the University of Wisconsin.

Besides the newness of the institutions and the opportunities for capable people, most of the settlers of midwestern states had an interest in and a respect for education that led to the rapid development of their institutions of higher learning. In contrast to the antiintellectualism of the South and the conservatism of the East, the societies of these new states experienced a ferment of new ideas and approaches.[4] Most important for this story, they emphasized an interaction between academic research and the solving of applied problems, an emphasis that meant that the emerging sciences of ecology and limnology had practical challenges to solve, that students and professors had opportunities for employment in applied ecology, and that ecologists could be coupled to the practical needs of their societies. These conditions created vigor in both ecology and limnology.

In northern Europe in a similar landscape, where glacial lakes were abundant, German limnology also developed actively. This growth was part of the rapid expansion of German science and technology in the nineteenth and early twentieth century, which was a result of the creation of the German empire from the German states and the industrialization of the empire. According to William Carr (1969), Germany was the leading industrial nation in Europe by 1900. After 1870, the emphasis of industrial development shifted from coal, iron, and heavy engineering to steel, chemicals, electrical engineering, and shipping. By 1913 Germany's share of world trade nearly equaled that of Britain and was twice that of France. This development was due partly to factors such as rapid population growth, high mobility from rural to urban regions, abundant raw materials, adequate bank credit, assistance from the state in the form of tariffs and subsidies, excellent management, and a skilled working class. In addition, German industrial development rested on scientific research and technological advances, including the invention of the electrical dynamo, synthetic dyes, and the process for making synthetic ammonia. Research, in turn, rested on an academic tradition in the many universities scattered throughout the German states and other German-speaking parts of Europe.

Established in 1891, the biological station at Plon, Holstein, northern Germany, became a center for German aquatic studies (Elster, 1974). August Thienemann (1882–1960) became the director of the Hydrobiologische Anstalt der Kaiser Wilhelm Gesellschaft, which replaced the station, in 1917. The influential technical journal *Archiv für Hydrobiologie* was published by the station beginning in 1916, and in 1925 Thienemann began another publishing program, *Der Binnengewasser Mitteleuropas.* These German activities were complemented by other work in Northern Europe, notably in Sweden, at Lund under Einer Naumann, and in Russia.

The significance of these North American and European limnological studies to ecosystem science stems from the fact that a lake is a relatively easily defined and discrete object. Thienemann and his colleagues made it clear that limnology studied lakes as whole systems in which the parts interacted to create the system. Although this concept is abstract on land, where the interactions are frequently separated in time and the vegetation dominates the system, in fresh water living organisms are relatively small in size and their life span short, so there is a much clearer linkage between living and physical-chemical processes.

The clarity of system boundaries and processes led Thienemann to a synthetic view of both limnology and ecology. In 1925 in *Der Binnengewasser der Mitteleuropas,* he stated:

> Here the biological and physiological sides of the conographic step toward uniformity come together in a real limnological synthesis. The whole life of waters is considered; the development of life in the water is conceived as a limnological unit of higher order. These units come about through a mutual exchange between biotope and biocoenosis, both standing in functional relationships, one to the other. The character of the living community is limited by its habitat, but it can also change through stability of the biotope, which may be rhythmical or cyclical so that the community returns to its normal condition only at intervals or of long duration when the community can persist.[5] . . .
>
> Each lake can be a living unity, therefore, each case is related to another. They are a microcosm, and therefore, this is the place to discover a higher structure of limnology, an organism of higher order whose organs interact in close mutual exchange. The lake is, we see, a limnological uniformity of the biotope and biocoenosis, of habitat and biotic community. This unity comes through the matter cycles in the relatively closed biotope of the lake. The question then is to characterize these uniformities of individual lakes by their species relationships and to group them together to form a system. These systems are self-standing in nature, considering all their essential characteristics. This shows that they are almost closed. (Thienemann, 1925, 187, 196)

In this article, Thienemann discussed several of the points I introduced in chapter 2. He was concerned about a synthetic view of a lake—realized through the interaction of the biota and its environment—as a system. Thienemann visualized this system as an organism of higher order, in the same way that Clements had viewed vegetation as a complex organism. In this excerpt, Thienemann refers to the biotic community as the *biocenosis,* a term invented by Karl Möbius, a professor of zoology at Kiel, Germany, in 1877, for the community of organisms growing on an oyster reef. Thienemann's term *biotope* refers to the environmental factors associated with the habitat of the biocenosis.

Thienemann's limnological experience led him in 1939 to express his concept of ecology in sixty numbered paragraphs in "Characteristics of General Ecology" (Grundzuge einer allgemeinen Ökologie). In this important article, he developed a clear statement of the interaction between living and nonliving elements that creates something larger or of higher order: "All separate factors in a biotope interact within the whole life space through mutual exchange. Here their totality works as an organizing factor—a local unit factor for a single biotope (Friederichs), or as a holocoen (Friederichs) for the total living space (or connected world)" (para. 18). In this quotation he used Friederichs's concept of holocoen to express the idea of ecosystem.[6] *Holocoen* and *ecosystem* are synonyms.

Thienemann was conscious of the role of chance in these systems and the flexible and variable nature of equilibrium:

> 30. The basic fact of the biocenotic equilibrium shows us clearly what the true nature of the biocoenosis is. One understands the term "life unit" to mean every life system in which organisms exist. This means that the biocoenosis is the higher order life unit for the single organism. If you call the life unit an organism of higher order, the purpose is only to draw a picture of it (to form a concept of it). Friederichs calls such biological whole bodies in which only single properties of an organism can be measured, biological "organizations," in analogy with organisms, but he stresses the point that there is no identity between them.
>
> In this sense the biocoenosis is an organization. Up to this point of time the organizations are not organisms, but the comparison is very helpful for science and research (Friederichs).
>
> The community is not only an aggregate or sum of organisms (and external factors coexisting in the same habitat) but a whole or system of organisms. The properties of the whole are not equal to the properties of its members. The single members retain properties of the whole which they do not lose if they become detached from the whole (Alverdes).

These conceptual statements place Thienemann in a group of holistic philosophers such as Smuts, Whitehead, and Meyer-Abich and the biologists Clements, Phillips, Friederichs, Weber, and Woltereck.[7] Elster (1974) in his history of limnology, commented that these ideas of Thienemann echoed Goethe[8] and other earlier cultural figures of Germany. In other words, we might label Thienemann, like Clements, a holistic materialist. He reasoned from his scientific studies of individual lakes toward an abstraction, the whole lake system.

Thienemann's ideas, however, were rejected totally by the empiricists, such as the leading experimental biologist, Max Hartmann, who in his influential book *Allgemeine Biologie* did not even mention ecology. Limnologists, just as plant ecologists, were having difficulty in defining and describing ecological systems in ways that linked the processes in them with the processes studied by physiologists and geneticists. Reductionistic studies in ecology, being based within biology, almost always focused on the finer description of species abundance and distribution within the habitat. Ecological reduction depended largely on taxonomic skills, and those demands almost always resulted in narrowing the field of study, leading the ecologist away from a broad understanding of ecological systems. The compilation of lists of the presence and abundance of species was not considered modern biology. Yet where the limnologist and ecologist could link through chemical processes, the biota, and the environment, a synthetic perspective was possible. Successful synthesis, however, had its own temptation, leading to speculation about the hierarchies of systems. When these abstractions lost their connection to the observational facts, they were also viewed critically by conventional biologists, such as Hartmann. Thus, ecology and limnology were positioned on an arête between two dangerous canyons—one dangerous because it led to an interminable argument over the taxonomic status and abundance of species, the other equally dangerous because it might lead to arid speculation about abstractions that could not be tested by observation or experimentation. These dangers were present everywhere but in Germany, where, possibly more than in other European countries, there were strong cultural supports for the holistic, integrative style of thinking expressed by limnologists such as Thienemann.

The Study of Lakes

Lake biologists began their studies with inventories of the biological organisms in lake water, on shore, and on lake bottoms. The operation was not a simple one, because many of the organisms were microscopic and special sampling methods had to be invented to sample aquatic habitats from boats, piers, and floating platforms. Victor Hensen (1887), an oceanographer, introduced the word *plankton* to refer to the smaller organisms in water and also devised quantitative methods to study them (Elster, 1974). A plankton net pulled through the water behind a boat collected a variety of animals and sampled a defined volume of water.

Of course, fish had been netted for centuries by people fishing and commercial methods of collecting fish needed only standardization and evaluation to be used to sample this part of the lake biota. The study of the fauna of the sediments was another matter entirely. As mentioned, Möbius had described the bottom organisms on an oyster reef as a biocenosis,[9] and his ideas were readily extended to lakes. The problem, however, was the collection of consistent samples of a bottom made of hard sediments or rocks. Eventually, effective dredges were constructed. Thienemann focused on the fauna of the bottom or the profundal zone, on what is now called the *benthic fauna*. In these structural studies the emphasis was on describing the numerical abundance of the species and changes in their distribution and abundance over time.

Besides descriptive studies of lake biology, the early limnologists were also concerned with describing the lake environment and reasoning from environmental change to biological change. Hermann Weber (1939a, 1939b) had developed his concept of *Umwelt*, or environment, in general biology as meaning not just the interaction between the organism and an environmental stimulus but as including all the factors necessary for the maintenance of the organism. The limnologists approached environment in this broad way. For example, patterns of water temperature with water depth were known at about the turn of the century. Birge and his associate Chancey Juday recognized that temperature caused the stratification of the water into layers, and that these layers were a central feature controlling the distribution and abundance of lake biota. Birge coined the terms *epilimnion* and *hypolimnion* to refer to the upper and lower layers of lake water separated by a layer where the temperature changed rapidly, called the *thermocline* (Birge, 1910).

Temperature stratification interacted with lake chemistry, especially with the dynamics of oxygen. Thienemann, Birge, Juday, and others showed that the distribution of organisms was related to the amount of oxygen in the water (Birge and Juday, 1911). In the profundal zone, after the lake stratifies, the living organisms are dependent upon the oxygen within the hypolimnion. As that oxygen is consumed in metabolism, anaerobic conditions may develop and the living organisms that require rich supplies of oxygen (such as fish) can no longer survive. Organic matter also requires oxygen for its decomposition, and large quantities of decomposing organic material can also deplete the oxygen. These various observations led Thienemann (1925) and Naumann (1932) to develop systems of lake classification, in which nutrient rich lakes and nutrient poor lakes formed ends of a continuum of lake types.

Lake Function

The structural approach to lake ecology satisfied the classical botanist and zoologist in focusing on the collection, identification, and counting of species, but it did not lead to a understanding of the lake as a system. This step required a study of the internal dynamic properties of the lake, what we collectively call its "function." As noted, the functional conception was present in Forbes's concept of the lake as a microcosm, as well as in Möbius's concept of the biocenosis. Nonetheless, it was difficult to implement these concepts until adequate technology was developed to enable the scientist to collect samples at a desired location repeatedly, thereby accumulating enough biological, physical, and chemical data to make functional comparisons and formulate and test hypotheses.

Forbes's (1887) point that the lake system is characterized by the cycling of matter and the building up and tearing down of life through the struggle for existence presented the essence of an agenda for the study of lake function. Cycling and building up and tearing down are different sides of the same coin. Building up protoplasm requires energy to drive the process and chemical materials to build the tissues; tearing down releases energy and materials. If one thinks of the living part of the system as linked intimately with the nonliving, then the cycle is the link between them through which materials flow. The problems with implementing this simple idea were that there are many different materials required by living organisms, the chemical transformations are very complex and in many cases poorly known, and life in a single lake is distributed in thousands of species and in millions of individuals. Further, the environmental sources for materials are complicated. In a lake the sources involve the water and sediments distributed in a variety of zones and layers.

One way through this tangle was to focus on a process central to understanding the dynamics of the entire lake. The process chosen was productivity. Not all ecologists agree on the definition of the terms *production, productivity,* and *product*. At a formative meeting of the International Biological Program in Poland in 1966, I recall three long evening discussions by an international group of ecologists where the meanings of these terms were debated from the viewpoint of different national languages. In Russian, productivity implies fruitfulness or potential and production implies the increase in the biomass over a given interval of time (Ivlev, 1945). Product is then the final phase of a given process. In English, productivity is the rate process, production is the general process, and product is the yield. In scientific reports written in the English language, the latter definitions have been used most commonly.

The production biology approach also made possible a quantitative de-

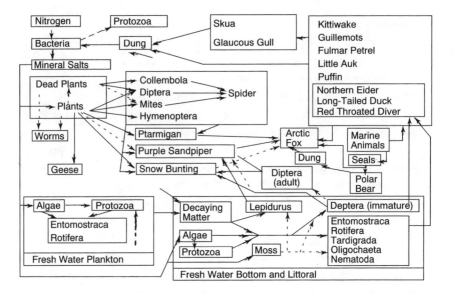

3.1 The nitrogen cycle on the arctic island, Bear Island, showing the web of interconnected organisms through feeding relationships (Summerhayes and Elton, 1923)

scription of transfers through the food web of the system, an idea grasped by naturalists as far back as Gilbert White of Selborn in 1789. Individuals in the biocenosis feed upon one another in intricate relationships, which, if drawn on a map, convey the impression of a web (fig. 3.1). The web metaphor implies that there is a symmetric order to feeding relationships. Charles Elton, who presented these ideas in a particularly clear way in 1927, used the terms *food chain* and *food cycle* to express them.[10]

In production biology it is necessary to determine the number of organisms in an area, their rates of reproduction and growth, their food requirements and the amount of food consumed, their excretion rates, metabolic rates, and the rate of death. With these data, one can develop a detailed cost accounting for the flow of energy or a chemical element into and out of a population. Richard Weigert pointed out in his 1976 survey of the history of ecological energetics that studies of this type were made as early as 1920 by Eikiti Hiratsuka for the silkworm in Japan. In a field situation it is more difficult to measure each parameter directly, but it can be done. For example, in studying the profundal of Lake Beloie, USSR, in 1939, E. V. Borutsky determined the growth and abundance of each instar of the dominant species, egg laying, emergence, migration in the lake, and losses through predation and death.

These data were put into a diagram showing the biomass dry weight changes for different lake depths during one annual cycle. The species were summed to give a picture of the dynamic behavior of the entire profundal zone.

This type of study is exceptionally detailed, and the data requirements quickly become insurmountable. In this age of computers, it is difficult to appreciate the problems posed when ecologists were obliged to account for the daily changes of individuals of myriad species on sheets of paper, using a pencil and slide rule. The studies such as those of Borutsky, Thienemann, Naumann, Birge, and Juday are astonishing in the amount of detailed handwork they required. The twin forces of difficult data manipulation and the urge to see the whole system led limnologists to group organisms into categories.

Actually, this step was logical as well as methodological because in many groups there is one dominant organism with many rare or subordinate species associated with it, and the dominant species can be used as a surrogate for the others. A simple way to organize these relationships was recognized early by naturalists. Plants were distinguished from animals, and animals were subdivided into herbivores and predators. Elton organized his food chains into a pyramid of numbers partly on this division. He said, "The small herbivorous animals which form the key-industries in the community are able to increase at a very high rate (chiefly by virtue of their small size) and are therefore able to provide a large margin of numbers over and above that which would be necessary to maintain their population in the absence of enemies. This margin supports the carnivores, which are larger in size and fewer in numbers" (Elton, 1939, 69). A few years later, Thienemann (1939) in *General Ecology* added one more category, "The three large groups of living organisms are the producers— the green plants, that use sun energy to build organic from inorganic substances—, the consumer—the animals that build their bodies from organic substances and consume to maintain their life-, and the reducers—the bacteria (and fungi) that are mineralizers, reducing the complex organic matter into the elementary composition" (Thienemann, 1939, 268–69).

The structured division into producers, consumers—which are subdivided into herbivores and carnivores—and reducers gives a simple organization to the complex biocenosis. For some, such as the Russian V. S. Ivlev (1945), this classification was too simple. He complained that the peculiar and important feeding habits of species and individuals were ignored by grouping organisms into a few categories. This criticism has been echoed by biologists ever since, but the seductively simple plans of Elton, Thienemann, and others and the urge to make the ecosystem concept quantitative led ecologists to overlook or ignore it.

Equilibrium

Studies of complex structures and functions would also be made easier if the patterns of nature were regular or stabile. *Equilibrium, balance,* and *stability* are all terms that imply constancy over time, predictability, and "unchangingness." The clocklike universe of the eighteenth-century imagination had this sort of predictability in it. It was thought that the planets moved in unchanging orbits and that nature was organized in a great chain of being that defined the place of every living creature. The only problem with this model is that our own experience shows that nature is constantly changing. We are surrounded by changing seasons, growing and developing plants and animals, and we age and become more experienced ourselves. Nevertheless, science in the nineteenth century seemed to support the idea that there was an underlying stability in the universe. Physicists showed that energy transformations followed laws that permitted determination of the thermal and mechanical properties of a body of uniform composition. Later, Tansley advanced his ecosystem concept based on his interpretation of these physical equilibria.

In 1878 the American physicist Josiah Willard Gibbs published his epochal monograph "On the Equilibrium of Heterogeneous Substances," which opened up the field of chemical equilibria and established the field of physical chemistry. Scientists grew increasing aware of the operational analogies between organic and inorganic matter; the difference between living and nonliving matter was that living organisms were in a state of dynamic equilibrium and nonliving matter was in a state of static equilibrium. The French physiologist Claude Bernard founded the science of physiology on a theory of equilibrium—the idea of the constancy of the internal environment of the body. Walter B. Cannon later coined the word *homeostasis* for this characteristic stability. It is not surprising that ecologists such as Clements and Forbes adapted these ideas and applied them to the objects they studied. The same process was occurring in the social sciences (Russett, 1966).

The problem for the student of lakes was to demonstrate a condition of equilibrium. Here there was a central problem in ecology, a problem I call a "confusion of levels." If we consider a lake as an object of interest, then the equilibrium of the lake involves a dynamic condition of its behavior. Thinking analogically from a individual human being, we would look for constancy in its health or well-being, its work capacity, and the stability of its personality. What is the health, work, and personality of a lake? Since most ecologists are trained as biologists, the answer to this question is almost always framed in terms of the biological elements of the system. These elements are the most dynamic and are responsive in a way different from the physical-chemical factors of the environment. Yet, the system is the interaction of living and nonliving matter within the lake basin. One way the behavior of a lake may be defined is through its conversion of input water from streams and overland flow into the output water of its draining river. This definition emphasizes the point that the behavior of the system reflects the system as a whole, not just a single component. We confuse levels if we judge the performance of the whole—the lake—solely by the performance of a component—the biota.

In this period, ecologists could do little more than advance the hypotheses that there were systems and that they displayed equilibrium conditions. It was not until many years later that it was possible to describe the behavior of systems directly and determine if the behavior was constant over time.

Thus, Forbes's assertions about lakes as microcosms were only partly capable of being verified. By the 1940s, however, biologists were building an understanding of the dynamic behavior of parts of lakes: the profundal, the plankton, the fish, and the littoral plants. Could these partial studies be put together to form a functional description of a whole lake?

THE LAKE AS A SYSTEM

Juday and Lake Mendota

Chancey Juday (1871–1944), professor at the University of Wisconsin, was among the first to attempt a functional description of a whole lake. His summary was based on extensive research by a team of investigators at the university who, with Birge, had studied Lake Mendota and other Wisconsin lakes since 1895. Juday (1940, 439) chose to present his whole lake assessment in the form of an energy budget. He said that "the annual energy budget of a lake may be regarded as comprising the energy received from the sun and sky each year and the expenditures or uses which the lake makes of this annual income of radiation. In general the annual income and outgo substantially balance each other." This idea was an exceptionally important one because it gave ecologists a theoretical structure that would show if an energy analysis was complete. According to the laws of thermodynamics, energy input and output must balance. If they do not, it means that some processes were missed in the analysis or that the measurements were faulty. Juday considered the physical heat budget of the lake, through which most energy flows, as separate from the biological budget. The physical heat budget involved solar radiation, melting ice, the heat produced by the water and the bottom, evaporation, and reflection. The biological budget involved the energy converted by photosynthesis and its use by each organism.

These Wisconsin scientists had data collected month by month, year by year, at a level of detail unlike that of any other group. For the biological budget, it was necessary to have information on the energy uptake and loss of the plankton, bottom flora and fauna, and the fish. Juday recognized the existence of food chains, where energy was transferred from aquatic plants to a series of animal populations, but he did not develop this picture. Rather, he limited himself to a comparison of the energy conversion through photosynthesis and its transfer to animals. Even so, Juday found it necessary to make several extrapolations and assumptions. He did not know the production of the plankton, which consisted of many different populations with different life cycles. Therefore he estimated the annual production of the plankton from an average turnover rate of the organic matter in the mean standing crop of plankton. Juday's willingness to group data for species and to extrapolate from these data to the whole system anticipated an approach common in later studies.

The main point drawn from this attempt at system analysis was that the energy flowing through the biota was an extremely small amount of the total budget. Juday found that the biota used about 1 percent of the total input. Of the energy collected through photosynthesis, about 5 grams of plant food were required to produce 1 gram of animal. The major part of the energy budget was in the physical processes of the lake, a subject in which the Wisconsin researchers had particular interest. These physical processes caused the lake to "take a deep breath" at the time of spring and fall overturn when all the waters of the lake were mixed. Birge (1907) had earlier used the metaphor "the lake respires," to draw attention to this phenomenon.

It is doubtful that Juday's attempt at total system description was satisfying to many biologists of the day. The role of the biota was insignificant in terms of system energetics, and it was necessary to make too many assumptions and ignore too many observable behaviors of the biota to calculate the energy intake and loss of groups of organisms, such as plankton. Yet, it is important to note that Juday did not use a holistic philosophy to justify his approach. The lake as an object was an adequate basis for his study. Forty or fifty years of limmnological research had established a foundation for his effort to bring together physical, chemical, and biological studies into a single description of a lake. This was a major advance in ecology, though Juday's effort was soon overshadowed by that of Raymond Lindeman.

Lindeman's Trophic-Dynamic Concept

The story of Raymond Lindeman (1915–42) is documented in a biography by Robert Cook (1977) and a memoir by Charles Reif (1986). He is well known to ecologists because his article "The Trophic Dynamic Aspect of Ecology"

3.2 Raymond Laurel Lindeman at the time he was a graduate student. Photograph provided by Charles Reif

(1942) is an acknowledged classic in the field, mentioned in all the major textbooks and regularly included in the lists of the literature in the field. Lindeman earned his doctorate in the department of zoology at the University of Minnesota, where his professors included Samuel Eddy (zoology) and William S. Cooper (botany), both key figures in American ecology. After graduation, with a Sterling Fellowship from Yale University, he worked with the distinguished limnologist and ecologist G. Evelyn Hutchinson.

Of the six monographs that constitute the Lindeman oeuvre, two are relevant to our story. These are "Seasonal Food Dynamics in a Senescent Lake" (1941b) and "The Trophic-Dynamic Aspect of Ecology," the result of five years of fieldwork on small Cedar Bog Lake near the University of Minnesota. Lindeman and his wife, Eleanor Hall Lindeman, sampled the biota of the lake with plankton nets and bottom dredge throughout the different seasons, as well as the water and bottom sediments, observing the growth, development, and distribution of the littoral vegetation, vertebrate animals, and the context of the lake in the forest. The study was unusual because it attempted to sample all elements of the biota at the same period of time. In a sense, it replicated the much larger team efforts on Lake Mendota and other Wisconsin lakes by Juday and Birge and on German and Swedish lakes by Thienemann and Naumann. It was necessary that the lake be small and well defined in order for two people to study the whole system. Cedar Bog Lake at the time of Lindeman's study had a depth of 1 meter, an area of 14,480 square meters, and a shoreline of 500 meters.

Lindeman concluded that the lake was an ecosystem. He was the first to implement Tansley's concept explicitly in a quantitative effort to define the system and describe and understand its dynamic behavior. The initial task Lindeman had in implementing the ecosystem concept was to organize an immense amount of data on the biology, distribution, and abundance of the lake plants and animals into a scheme or pattern. His approach, evolved from that of others, was a unique and creative one called the *trophic dynamic aspect*. The first stage in building the trophic-dynamic approach was derived from prior studies by Möbius (1877), Forbes (1887), Forel (1907), Shelford (1918), Alsterberg (1922, 1925), Thienemann (1926), and Strom (1928), who had suggested that the biota could be described as a network of interactions with groups of organisms linked by feeding. Lindeman created his scheme to emphasize the special character of Cedar Bog Lake. He placed *ooze,* or the material that filled the bottom of the lake at the center of the diagram, surrounded by two different flow patterns (fig. 3.3). One set of linkages was based on plankton and the other on littoral pond weeds. All parts of the biota were linked to the ooze. By constructing the diagram in this way, Lindeman emphasized the interaction of the living and nonliving parts of the lake, which were intimately interconnected. Lindeman's diagram was similar, in his emphasis on the ooze, to those of Thienemann (1926), Strom (1928), Rawson (1930), and Wasmund (1930), all of which were illustrated and discussed in his thesis. We can assume that his diagram was a central element for his organization of information on Cedar Bog Lake, since he published the diagram in each of his major publications.

His second task involved formulating a way to go beyond earlier attempts to describe lake metabolism and energy flow, which Lindeman called the *dynamic species distribution approach*. He solved this problem by focusing on the food cycle within the ecosystem, whereby he could link the living and nonliving parts of the system and organize species into groups based on their food habits. He identified species, determined their food habits from observation, experiment, or the literature, and then organized the various species into food groups. The food groups became the cells of his ecosystem diagram. Feeding

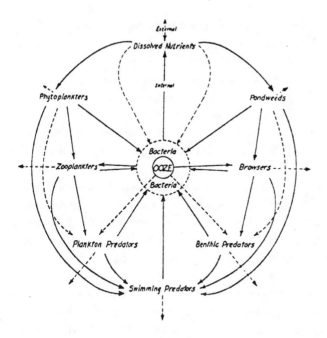

3.3 Lindeman's diagram of a food cycle in Cedar Bog Lake (Lindeman, 1941b). Published with permission of the *American Midland Naturalist*

was made comparable between diverse biotic groups by converting food into energy units.

Finally, Lindeman used this new trophic-dynamic approach to understand ecological succession. The successional paradigm must have been of special interest to him because one of his professors, William Cooper, was a major contributor to the scientific study of successional.[11] Cedar Bog Lake was a senescent lake, formed by a melting ice block after the retreat of glaciation, which by the 1940s had reached a point where it was nearly filled with peat and marl. The lake was primarily understood as a stage in lake succession, an imbalance between the forces of production and catabolism. Plant organic material gradually increased and changed the depth and quality of the lake. Lindeman's aim was to use the trophic-dynamic approach to understand the balance between these processes.

In 1939, Lindeman met Edward S. Deevey, a former student of Hutchinson, at a hydrobiology meeting at Columbus, Ohio, and found that they had common interests. Deevey had studied the biological history of lakes by sediment core analysis, and he and Hutchinson were interested in lake succession. It was Deevey's suggestion that led Lindeman to apply for a fellowship at Yale to work with Hutchinson following his graduation in March 1941. The move was important in his interpretation of the data he and his wife had collected on Cedar Bog Lake. In the meantime, his article on the seasonal dynamics of a senescent lake, which contained most of the biological data from his thesis, was accepted for publication in the *American Midland Naturalist*.

The major portion of the seasonal dynamics treatise recounts the abundance, biomass, and feeding relationships of the biota, organized into the categories of the food cycle diagram. The dominant species in each group were reported in detail and in some cases were surrogates for other, less abundant or less well known species. Lindeman found great variation among these species. From one year to the next the dominant groups or the species within a group differed widely. He explained this variation as being caused partly by factors external to the lake, such as water inflow, rainfall, winter temperature, ice cover, and chemical flux.

He moved beyond his thesis into three new perspectives. First, the biomass of the species were converted to energy units, using the conversion factors determined by Birge and Juday for Lake Mendota.[12] In the thesis, the figure that reported the biomass dynamics of food groups was converted to energy by altering the abscissa of the figure, using the conversion factor of 10 grams centrifuged wet weight per square meter as equal to one calorie per square centimeter. Second, the annual production of the food groups, calculated by assuming appropriate turnover rates for each group—following the approach of Juday for Lake Mendota—were summed and presented as three trophic groups: producers, primary consumers, and secondary consumers. Third, efficiencies of transfer between groups were calculated by comparing the amount of energy in annual production in various food and trophic groups. For example, about 10 percent of the production of the producers was in the production of primary consumers, and about 19 percent of that in secondary consumers compared to that of the primary consumers.

These important advances in theory were a natural extension of Lindeman's thesis, obtained by applying an energetic approach to biomass, following the lead of Juday. Energy was a key part of metabolic physiology, recognized as central by other ecologists, but not yet readily applied to the metabolism of ecological systems. Lindeman had more detailed and concentrated information than Juday. By applying energetics to his standing crop data, he could extend Juday's approach into a more detailed description and expansion of the concept of ratios or efficiencies. Possibly, there was also encouragement from Hutchinson and Deevey,[13] who used ratios of species remains in sediment cores to demonstrate patterns of change over depth.

In his thesis, the energetic approach was based partly on an article by Edward Haskell (1940), in which Haskell attempted to create a physical mathematical theory of ecology. Haskell characterized the environment as entropic in nature, plants as regions of integrated entropic and biocatalytic processes, and animals as entropic, biocatalytic, and signalloid processes, and proposed that each might be described by differential equations of sequentially higher order. Hutchinson (Cook, 1977) said that he did not believe half of Haskell's propositions, and Lindeman did not reference Haskell in his published papers, since the theoretical underpinning for trophic dynamics was no longer necessary.

The organization of the food groups recognized in his trophic diagram into trophic levels was also a natural step, since Thienemann (1926) and other limnologists had used categories of producers, consumers, and reducers to organize biological data. Yet Lindeman was faced with an exceptionally varied fauna and flora that changed dominance each year. It was easier to organize these alternating species into groups that represented their collective feeding relationships than to keep them separate as biological units. In this way he could discuss components of ecosystems that had continuity year to year and season to season.

"The Trophic Dynamic Aspect of Ecology" was based on the final chapter in his thesis, "Cedar Bog Lake: The Ecosystem or the Trophic Dynamic-Viewpoint in Ecology." The article was started in February 1941 but after he moved to Yale in September, it became dramatically altered, incorporating ideas provided by Hutchinson, who had arrived at some of Lindeman's conclusions independently.[14] Still, this draft was rejected when it was submitted to *Ecology*. Cook (1977) tells us that the referees were two of the most distinguished American limnologists, Juday and Paul Welch of the University of Michigan. Both based their rejection on the theoretical nature of the article, asserting that it was premature and based on too particular a lake.[15] The reviewers also thought that a theoretical article was inappropriate for *Ecology*. Although it seems ironic that Juday, who had published on the energetics of Lake Mendota several years before, would reject Lindeman's article, he and Birge were studying 529 lakes in northeastern Wisconsin. Their survey showed how special each lake was and how difficult it was to demonstrate a single-property characteristic for all lakes. Hutchinson appealed to *Ecology*'s editor, Thomas Park, of Chicago, strongly supporting the need for theoretical monographs in the journal of the Ecological Society of America and taking responsibility for some of the ideas expressed in the article.[16] Park offered to consider a further revision. When the fourth draft was submitted in March 1942, Park sent it to another distinguished ecologist, W. C. Allee, of Chicago, who gave it lukewarm approval. Park, on his own decision, accepted the article, and it was published in October 1942, after Raymond Lindeman had died. The article was published with a generous addendum by Hutchinson.[17]

A comparison of the finished article to the final thesis chapter shows how far Lindeman's thinking had advanced. The format of both are similar, with sections on definitions and concepts, followed by the dynamics of ecosystems, and concluding with the application of trophic dynamics to ecological succession. The concept section in his article asserted that the ecosystem was the key concept to understanding Cedar Bog Lake because it involved the interaction of living and nonliving parts of the system. The ecosystem dynamics section in his thesis was, as mentioned earlier, based on Haskell's (1940) idea of extending physical and mathematical ideas to ecology. Lindeman admitted that he did not know how to implement Haskell's concepts. Instead, he turned to productivity (defined as the amount produced per unit of time) to organize the static data on species biomass and numbers and attempted to explain a change in productivity by the change in the nutrient supply required to build the biomass and by external factors, such as rainfall that raised or lowered the lake level. In the article, Haskell's ideas were dropped in favor of a symbology derived from Hutchinson, where each trophic group was represented by the difference in the rate of energy entering or leaving the group. In this scheme, productivity was defined as the transfer between trophic groups, with emphasis on comparisons using ratios and efficiencies. Productivity was corrected to account for losses from metabolism, predation, and decomposition. Lindeman found a progressive change in efficiency up or down the trophic sequence. For example, the efficiency of transfer appears to increase as one moves from producers to tertiary consumers. This section in the treatise is so fundamentally changed from Lindeman's thesis that it is entirely new, except for carrying over the concept of trophic groups or levels.

The final portion of his thesis represented his argument about understanding trophic dynamics in the context of ecological succession, which shifted in the article to a focus on trophic dynamics to explain succession as ecosystem development. Lindeman accepted conventional theories of ecological succession in general, and of lake succession in particular. These included the theories of Clements that succession is a process leading to an equilibrium state set by the regional climate and of Thienemann and other limnologists that lake succession is a process that begins with oligotrophy, where production is limited by a low nutrient supply, progresses to a eutrophic state—where production operates at a high level of activity—and eventually, as the lake fills in, becomes a terrestrial bog and finally a forest.

When Lindeman compared his data from Cedar Bog Lake with those for other lakes, for example, the energy production with that of Lake Mendota and

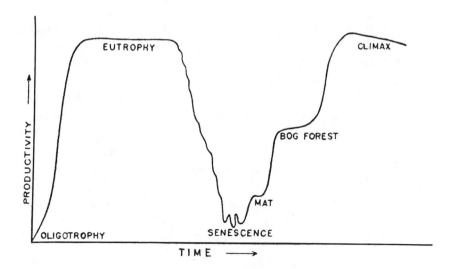

3.4 Lindeman's hypothetical relationship of productivity changes over time in a deep lake, which eventually, following the theory of Clements, would become a climax forest (Lindeman, 1942, 413). Published with permission of the Ecological Society of America

the process of deposition with that of Linsley Pond, Connecticut, he found that Cedar Bog Lake was not, as he had earlier thought and remarked on in the thesis, highly productive. Lake Mendota producers were about four times as productive as those in Cedar Bog Lake. In his article, Lindeman separated Cedar Bog Lake from other lakes and drew a new curve of productivity and time (fig. 3.4), showing a long period of stable productivity, followed by a decline to senescence, then an increase as terrestrial systems developed and covered the former lake bed.

In his summary, two conclusions came from his thesis and three were new. The two from the thesis were that the ecosystem was the fundamental unit of trophic-dynamics and that the pond can be organized into trophic levels. A further conclusion was derived from the thesis but was not stated explicitly. This conclusion was that the more remote the level was from the source of energy, the less dependent it was on one source. The new concepts stated that the energy of the contributing level *a* is greater than that of the receiving level *b,* that respiratory loss is higher at the higher trophic levels, and that consumers at the higher levels are more efficient in energy transfer than those at lower levels. His two final conclusions referred to succession, that productivity and efficiency increase in early succession and that consumer efficiencies increase during succession.

Lindeman's trophic dynamic treatise implemented the ecosystem concept. Further, it showed that a single individual, with a helper, could study an entire ecosystem and describe the activities of the biota in an acceptable fashion. It was within the ability of a single investigator to study a small lake, spring, or pond,[18] while a team was required to study a large system, such as Lake Mendota. From a theoretical view, it demonstrated that one could identify an object in a physical environment, defined by the morphometric boundaries of the depression, and study it as a system. This system had a structure of living species and nonliving material. It also had a metabolism and converted input energy from the sun into heat energy or energy stored in sediments and ooze. It received nutrients from its basin, recycled them through the biota and nonliving materials, and exported them in water, atmospheric gases, pupating insects, and food for terrestrial animals. Finally, the lake ecosystem had a development pattern. It changed through time. This change, called succession, was a function of the interactions of the biota and nonliving material with the external climate, nutrients, and human disturbance. Lindeman focused on the first and last point; he did not develop the theme of nutrient cycles. To complete the story I briefly consider this theme.

A Supporting Theme

Lindeman introduced the concept of the food cycle in his writings. The phrase was taken directly from Thienemann's *Der Nahrungskreislauf,* but Thienemann did not develop or fully implement the concept in its trophic-dynamic aspect. The concept includes the idea that nutrients enter and leave a system through transport mechanisms in the atmosphere, hydrosphere, and biosphere, and that living organisms within the system, which are actually subcomponents of the ecosystem, take up, store, and release nutrient elements to their environment. In this way the living part is coupled to the nonliving part of the system, which functions alternately as a source or sink of elements.

The concept of nutrient cycling enters ecology from several sources. The first is geochemistry, which focuses on the chemical flows between the biota, atmosphere, hydrosphere, and lithosphere (Fortescue, 1992). Geochemical studies began about the same time as ecological work. The pioneer American geochemist F. W. Clarke had published a summary of geochemical data in *The Bulletin of the U.S. Geological Survey* in 1908. To understand the linkage of living and nonliving parts of the ecosystem, the chemistry of the environment had to be integrated with biology. This process of integration was undertaken by several individuals. The first was Vladimir Vernadsky (1863–1945), a Russian, who developed the concept of biogeochemistry. Vernadsky was basically a mineralogist (his doctoral thesis was "The Phenomena of Crazing in Crystalline Matter"), but he was active in many areas of science, science management, and social matters. During his stay at the Sorbonne in 1923–26, he wrote on geochemistry, mineralogy, crystallography, biochemistry, marine chemistry, the evolution of life, geochemical activity, and futurology (Baladin, 1982). Rudolf Baladin (1982, 41–42) comments, "His capacity for work was amazing. Up to his very old age he worked 10–12 hours a day, and sometimes even longer, combining a consistent sharp interest in investigations with the rigid organization of his work."

Vernadsky merged the data from several fields into one perspective that considered the earth as a chemical system where the elements cycled between the various parts. His perspective had a strong influence within the Soviet Union, where he was recognized as a scientific leader (he organized the first national scientific academy, the Ukrainian Academy of Science, and was its first president immediately after the civil war), and was also recognized throughout the world through his lectures, travels, and books. His book, *The Biosphere,* first published in 1926 in Leningrad and in 1929 in French as *La Biosphere,* had wide impact as a scientific expression of a global system of man and nature, which was an antidote to the virulent nationalism that was being expressed at the time, especially in Europe.

Vernadsky's student A. E. Fersman (1883–1945) formulated methods to map geochemical provinces, thus creating a spatial component in the field. Another pioneer was the Norwegian V. M. Goldschmidt (1888–1947), who emphasized the integration of geochemical cycles. All these pioneers were chemical holists in that they were concerned with the movements of all elements in the biota and environmental spheres.

Fortescue also discusses in his history of landscape geochemistry the Russian school of soil science and geography of V. V. Dokuchaev (1846–1903). Two of Dokuchaev's ideas that are relevant to the ecosystem concept are that every natural zone constitutes a regular, natural complex in which living and nonliving aspects of nature are closely associated, and that soil is an independent natural body that must not be mistaken for surface rocks. These ideas were at the heart of Dokuchaev's thinking and provided a base for the organization of the science of landscape geochemistry.

Another contributor to the concept of nutrient cycling was an American, Alfred Lotka (1880–1949). Lotka was born in Europe of American parents and educated in Germany, France, and England. He was a physical chemist and for much of his life worked in industry and government. His scientific work was done in his spare time, published in a variety of scientific journals, but was largely unrecognized. Sharon Kingsland (1985, 29) describes his work: "Lotka labored on, building his program piece by piece; but he was like a mole, anxious to soar yet equipped only to tunnel deeper underground. His schemes and scratches had brought him no closer to recognition by 1920, when his small movements caught the eye of an eagle, which swooped down to raise him up in its large but friendly talons. The eagle was Raymond Pearl of the Johns Hopkins University." Pearl was professor of biometry and vital statistics in the new School of Hygiene and Public Health, and he offered assistance to Lotka. Lotka moved to Baltimore in 1922 and began writing *Elements of Physical Biology*, in which he viewed the entire earth as a single system with the various components linked by exchanges of chemical elements and driven by solar energy. He commented, "We shall probably fare better if we constantly recall that the physical object before us is an undivided system, that the divisions we make therein are more or less arbitrary importations, psychological rather than physical, and as such, are likely to introduce complications into the expression of natural laws operating upon the system as a whole" (p. 158). Lotka anticipated the study of food chains, producers and consumers, cycles of water, nitrogen, carbon and other elements, and the mathematics of trophic transfer. Unfortunately, his book did not stimulate the creation of a new science of physical biology. On the one hand physicists, generally, were not interested in his ideas. On the other hand, some ecologists found the ideas congenial since his systems approach was based on a quantitative database from a wide variety of disciplines, and he implemented at a global level what many, such as Thienemann, Birge, and Juday, were trying to do on a smaller scale. Charles Adams, president of the Ecological Society of America in 1923, recognized the worth of Lotka's book and encouraged him to write a review in *Ecology* (Adams, 1915). He never did, but he did join the society in 1925.

Lotka's book was reprinted in 1956 as *Elements in Mathematical Biology*, and it became an ecological classic that was widely consulted in the 1960s and 1970s. Although it appears to be a precursor to ecosystem ecology, we do not know how many ecologists were familiar with the first edition of Lotka's book. It is not cited in the literature that defines the developments we have discussed, but some ecologists, other than C. C. Adams, may have been influenced by it. Now, at the end of the twentieth century, when we are concerned about changes in the global climate, it has, like Vernadsky's writings, a modern feeling.

Ecological studies in biogeochemistry were advanced by the group around Hutchinson of Yale University. Hutchinson played a pivotal role in the story of ecosystem ecology, as well as in American ecology in general. He was born in England and grew up in the academic city of Cambridge. Later, he worked in South Africa before moving to Yale in 1928. Hutchinson was familiar with

Vernadsky's ideas, was a colleague of Vernadsky's son, who was a Yale professor, and translated several of Vernadsky's articles. In the 1930s, Hutchinson and his students began their studies of the element cycles in the small lake Linsley Pond on the outskirts of New Haven.

One of the organizing principles of Hutchinson's theoretical work was the familiar concept of system equilibrium or balance. He thought that systems evidenced processes of self-regulation that produced and maintained equilibrium conditions. As a case in point, Edward Deevey, in his examination of cores of the organic sediments that had accumulated in Linsley Pond since postglaciation, found a period of high productivity followed by a relatively constant rate. Hutchinson interpreted this pattern of productivity as due to modification of the environment by the pond organisms, and forming an ecology of self-regulation leading to equilibrium conditions. Hutchinson's other students examined biogeochemical cycles, productivity, and equilibrium in other systems. Among these was Howard T. Odum, whose thesis concerned the strontium cycle.

The second path to nutrient cycling came from physiology. The German agricultural chemist Justus Liebig is usually cited as the first scientist to recognize that chemical elements were limiting to plant growth (Liebig, 1876). Physiologists concerned about the nutrition of animals and humans or plant growth demonstrated that many of the chemical elements of the periodic table were essential for living organisms. A comparison of the geochemical abundances of elements in the biosphere does not correspond closely with the abundances in the human body (Lotka, 1925), leading to the conclusion that living systems evolved their own unique chemical compositions, which are maintained in a variety of chemical environments.

The selection of essential chemical elements from the environment and the interplay between anabolic and catabolic processes that maintain a steady state or homeostasis provided the ecologist with a conceptual framework that could be applied by analogy to ecological systems. Actual biogeochemical work was initiated by ecologists such as Thienemann and his group at Plön, and G. Evelyn Hutchinson[19] and his students at Yale University.

With Raymond Lindeman we see for the first time a deliberate effort to implement Tansley's ecosystem concept. His focus was on the dynamic processes of the ecosystem. The idiom in which he expressed this process was energy. Lindeman introduced most of the major questions and concepts of modern ecological energetics, including questions about the length of food chains, the efficiency of trophic transfers, the storage of energy at different levels, the rates of primary productivity, the problems of correcting energy values for losses due to respiration, predation, and decomposition, and the role of bacteria and microorganisms in cycling dead organic matter. In addition, he made clear the idea that ecosystems develop through ecological succession and are tied to the energy dynamics of the system and the concept that nutrient cycling, as food cycling, is linked to the wider biogeochemical cycles coupling one ecosystem with another.

Approximately seven years after the ecosystem concept was introduced, Lindeman had defined an outline for a research program that would occupy ecologists for the next forty or more years. This program asserted that nature is organized into ecological systems that are recognizable objects such as lakes and that have an origin and development leading to a steady state or dynamic equilibrium. These systems have a structure—defined as a network of feeding relationships among their species populations—that can be simplified by grouping the populations into food chains or trophic levels. The ecosystem has a behavior, beyond that of development over time, that involves the processing of energy received from the sun or other systems into heat and work, and the processing of chemical elements imported into the system into various storages and outputs. These are the energy flows and the nutrients cycling between the species populations and between them and the nonliving parts of the system. The structure and the function of the system can be expressed mathematically as a series of equations describing the interactions between system components.

This full program was described in rough terms only in the 1940s because of the difficulties encountered in the study of a whole system in the field and in analysis of the collected data without the help of a computer. Yet the skeleton was clear, it only remained to build up the flesh. This became a possibility after the hiatus imposed on research during the second world war.

CHAPTER 4

Transformation and
Development of the Concept

In the previous chapter I showed how Tansley's ecosystem concept was implemented by Raymond Lindeman in his study of Cedar Bog Lake in Minnesota. With this, Lindeman introduced another way to view the natural world. The species occurring in the community were identified and their living mass, or biomass, was determined. They then were organized into a model that showed how they interacted with each other—based on their feeding, predation, and parasitic relations—and with the environment. He assumed the species could be aggregated into groups, (called trophic levels) of species feeding directly on living plants, those feeding on dead organic material, and those feeding on herbivores. The amount of energy or food flowing from one trophic level to another, represented either as the amount stored in a group at one time or as rates of flow, could then be compared using ratios of efficiency.

Lindeman's use of energy to express the relations between trophic levels again linked ecology with thermodynamic theory in physics. Lotka, Haskell, and Juday had recognized this linkage earlier.[1] Lotka, in his book *Elements of Physical Biology,* on "Energy Transformers in Nature" (chap. 24), anticipated many of the applications used by Lindeman, including energy accumulation and dissipation at each step in the sequence of transfers, the coupling of plants and animals, and calculating the efficiency of the transformer. Lindeman's formulation also fit some of the concepts of economics, which like physics also had become well developed mathematically and theoretically, especially in econometrics.[2] As a parallel example between ecology and economics, energy could be considered a form of natural currency and the coupled populations to be engaged in a process of exchange. The energy laws permitted an ecologist to calculate a balance sheet where energy input equaled energy outgo. Howard Odum, a student of Hutchinson, developed this aspect of the formulation into a more elaborate energy theory (Odum, 1971).

In all of these subjects, phenomena represented by the words coupling, accumulation, transformation, and efficiency were being treated within a single, unified perspective. This perspective, labeled *a general systems approach* (Rashevsky, 1956; Bertalanffy, 1950), applied a new organic, holistic metaphor to nature. It demonstrated that all natural and human-built systems had a fundamentally similar structure and function. It was in sharp contrast to the organic metaphor of Clements and the theoretical concept of the "whole," the *holoceon* or the *biocenosis* of European ecologists and philosophers because it was physical and had a machinelike character. Indeed, at the time of the second world war, the application of systems engineering to weapon guidance became exceptionally successful and led to Norbert Wiener's "cybernetics" and to C. E. Shannon and W. Weaver's mathematical theory of information at Bell Laboratories. The application of these mechanistic approaches to the reconstruction of war-devastated areas and to the peaceful development of society was widely accepted. Tansley's ecosystem concept was preadapted to this period, but in the beginning it was poorly known and seldom used.

POPULARIZATION OF THE CONCEPT

The ecosystem concept became known outside of the limited literature of ecological science and its specialists when Eugene Odum (fig. 4.1) made it a central concept in his textbook *Fundamentals of Ecology* in 1953. Odum's text had several features that made it unusual. He placed the ecosystem and biogeochemical cycles near the beginning of the book and organized it in an easily understood way. For me, and I suspect for many others, the treatment of all of ecology in a concise format made Odum's text especially attractive and useful. The significance of his "top-down" approach became apparent only later.[3]

Fundamentals of Ecology was not the only textbook to examine the data and concepts of ecosystems as developed by Lindeman, Hutchinson, and others. Another important book was *Principles of Animal Ecology*, published in 1949. This text was more than its title implied. It took the middle ground between what its authors described as the encyclopedic treatise of present-day knowledge about the subject and a brief statement of underlying principles, together with a sampling of the evidence on which they were based. There were five distinguished authors, W. C. Allee, Alfred E. Emerson, Orlando Park, Thomas

4.1 Eugene P. Odum, 1970s. Photograph courtesy of the Institute of Ecology, University of Georgia

Park, and Karl P. Schmidt, all from institutions in the Chicago area. As might be expected, the book was written in unusual depth. As a consequence, it was a compendium of careful, considered, and full discussions of ecology, with an emphasis on animals. The oversized book was 834 pages long. Its excellence and thoroughness was undisputed.

A section of *Principles of Animal Ecology*, "The Community," discussed food webs, energy flow, and the efficiency of trophic transfers. Although the ecosystem concept was included in the book, it was most prominently treated in the sections on evolution and adaptation. At this time (1949), the ecosystem and community concepts were not clearly separate.

In large, the major community may be defined as a natural assemblage of organisms which, together with its habitat, has reached a survival level such that it is relatively independent of adjacent assemblages of equal rank; to this extent, given radiant energy, it is self sustaining.

The functional integrity of the community is a logical extension of the facts examined, since it becomes apparent that the community must be the natural unit of organization of ecology, and hence is the smallest such unit that is or can be self-sustaining, or is continuously sustained by inflow of food materials. It is composed of a variable number of species populations, which occupy continuous or discontinuous portions of the physico-biological environment, the habitat niches. (Allee et al. 1949, 436, 437)

The section "Community Organization: Metabolism" contained most of the relevant information about the ecosystem concept. It focused on the food web and considered two primary production processes. One of these involved the reorganization of matter by bacteria and the other, photosynthetic activity. Its consideration of catabolic processes led to a discussion of food chains, food webs, pyramids of numbers, and biomass. Energetics provided for the comparison of the efficiency of transfers between trophic levels.

Its account of the ecosystem, defined as community metabolism, was based not only on Lindeman's (1942) data from Cedar Bog Lake, but also Birge and Juday's (1934) work on Lake Mendota, George Clarke's (1946) studies of the George Bank in the north Atlantic Ocean, and Gordon Riley's (1941) studies of Long Island Sound. The data are brought together in a objective way, with no evidence of the holistic arguments that motivated ecologists studying lakes a few years before. It repeatedly states that community metabolism is the sum of the metabolism of the species populations that compose the community. Its focus is entirely on food transfers varying over space and time.

Principles of Animal Ecology is rooted in the authors' vast knowledge of species populations. The stratification of the community, along with its metabolism, periodicity, and evolution, is understood as being aggregations of stratification, metabolism, periodicity, and evolution of species populations. The community concept is built from an understanding of these populations. It explicitly stated that there were no whole community studies yet; indeed, there might never be, but that understanding the community is a synthetic process, building up knowledge from the components as in the construction of an engine. The contrast with Eugene Odum's top-down approach is obvious.

Another text appeared in 1952, the year prior to Odum's book. This was *Natural Communities,* written by the animal ecologist Lee R. Dice, a professor at the University of Michigan. Dice was a student of Joseph Grinnell of the University of California at Berkeley. He spent most of his professional career in Michigan as a mammalian population ecologist and geneticist.

Dice's text on the ecological community incorporated the experience of both plant and animal ecology in a modern synthesis. It began with a examination of the classification and nomenclature of communities and considered methods of studying communities. Dice thought that ecology studied three levels of complexity—the individual, species population, and the ecological community. Community ecology was grounded in population ecology, where Dice had unusual knowledge and experience, so many of the key ideas of population ecology were included in the text. These included fluctuations in populations, home range and territory, the effects of social behavior, and ecological relations between species. Dice also incorporated the new ideas of Lindeman, Hutchinson, and others in chapters on food relationships and equilibrium. The term *ecosystem* occurs frequently. A final chapter on philosophy, an unusual addition to a book on ecology, serves to consider the nature of plant community from the perspectives of Clements and Gleason.

Dice's text was unusual and distinct from a commercial product like Odum's book. It represented the mature experience of an ecologist with broad experience and interests. Although it covered the same ground as the Chicago school's animal ecology, it was an individual's synthesis of the material. Like its author, it was a gracious book, careful and quietly grounded in a statistical and philosophical understanding of a complex subject. It also represented an earlier era of academic publication. It was an artistic and literary product, as well as a scientific synthesis and textbook, and included many sketches by the University of Michigan Museum artist Carlton W. Angell.

Both *Principles of Animal Ecology* and *Natural Communities* represent steps in the transition from the traditional focus of ecology on natural history and the abundance of organisms to the mechanistic treatment of ecosystems. The concepts of Lindeman, Hutchinson, and other ecosystem pioneers were useful in describing the interaction between feeding populations and in advancing a more modern reason for natural stability and equilibrium. Yet the emphasis still lay with populations. Because each species population displays distinct ecological properties, a complete analysis required a discussion of each relevant property.

Eugene Odum's text was entirely different. It was short, simple, and emphasized principles. In each section he introduced a concept with a concise statement, explained the concept, and placed it in context, followed by one or more examples using scientific data. This approach was almost revolutionary. It was conceptually clear and permitted the material to be scanned quickly. Most important, it permitted Odum to highlight and emphasize principles and concepts that he felt were the fundamental elements of the science within a logical and coherent framework.

Fundamentals of Ecology was written partly as the consequence of an argument in the department of zoology at the University of Georgia. At that time the university was a small, southern institution that had just begun offering the doctorate degree in a few departments. Zoology, under the leadership of George Boyd, was one of the most scholarly and active departments of the university. Odum served as its ecologist. In the late 1940s the department of zoology began discussing a revision of its core teaching program, and Odum pressed for the inclusion of ecology. Although Odum was, and still is, an aggressive, combative proponent for his subject, he failed to convince his colleagues of the importance of ecology. *Fundamentals of Ecology* began as a reaction to this failure.[4]

The form of the book also had its origin in this disagreement. One of the criticisms against ecology by the zoologists was that it had no principles. Odum began compiling a list of the principles of ecology. This list was eventually organized into chapters and became the key structural element of his text. Important as this step was, however, it was less significant, in retrospect, than the order in which he chose to present his principles. *Fundamentals* begins with a short introduction that tells the reader what ecology is and how it is organized. Immediately following, in the first full-length presentation, Odum (1953, 9) explains the concept of ecosystem. He says that "any entity or natural unit that includes living and nonliving parts interacting to produce a stable system in which the exchange of materials between the living and nonliving parts follows circular paths is an ecological system or ecosystem. The ecosystem is the largest functional unit in ecology, since it includes both organisms (biotic communities) and abiotic environment, each influencing the properties of the other and both necessary for maintenance of life as we have it on the earth. A lake is an example of an ecosystem."

In his explanation of the ecosystem concept, Odum developed several distinct ideas. First, the largest ecosystem is the entire earth and the biosphere is that portion of the earth where ecosystems operate. Second, ecosystems may be of various sizes, from the biosphere to a pond. Third, animals and nongreen plants are dependent upon plants that manufacture protein, carbohydrates, and fats through photosynthesis; plants are controlled by animals, and both are influenced by bacteria. Fourth, organisms also influence the abiotic environment. Fifth, humans have the ability to drastically alter ecosystems. Thus, an understanding of the ecosystem concept and the realization that mankind is part of these complex biogeochemical cycles is fundamental to ecology and to human affairs generally. By page twelve of his textbook Odum had laid out an agenda that anticipated the direction taken by the ecological movement over the next twenty years.

Fundamentals unfolds from this beginning to consider, in turn, limiting factors, energetics, the food chain, pyramids of numbers, biomass, energy, productivity, and then principles at the level of the population, between two or more species and at the community level. The final third of the book is concerned with habitats, focusing on the differences between freshwater, marine and terrestrial environments, and applied ecology.[5]

It would be unfair to suggest that the origin of Odum's text was only a reaction to an academic disagreement. As early as 1945, he was gathering materials and ideas for two textbooks, one on ornithology and the other on general ecology.[6] He explained his interest in textbook writing to Lowell Noland at the University of Wisconsin by saying that he felt "rather strongly that the field of ecology has been badly presented and has broken down into too many antagonistic subdivisions. I'd like to try to help put ecology on a firm foundation such as is now enjoyed by genetics, for example." He did not say that textbook writing was a family tradition, but that would have been a perfectly valid explanation as well. Odum was born into an academic family. His father and mother were both well educated, and his father, Howard W. Odum, was a distinguished sociologist at the University of North Carolina, Chapel Hill. Howard W. Odum was the author of numerous books, including the significant *Southern Regions of the United States,* which "became a basic book in courses and seminars, had an important impact on policy and thought, went through four printings and had books, commentaries, and pamphlets written about it."[7] According to Eugene Odum, his father encouraged him when the attempt to bring the enormous number of facts and ideas encompassed by the word *ecology* into a manageable book seemed overwhelming to him.

In addition, his younger brother, Howard Thomas (Tom) Odum (fig. 4.2), played an important role in the development of the book. Tom Odum was a doctoral student studying under Hutchinson at the time. Eugene Odum credits Hutchinson's ideas, transmitted as copies of class notes by Tom Odum, as being a key inspiration in his using the ecosystem as a central organizing theme for *Fundamentals of Ecology.* Tom was an equally creative member of the Odum family and better trained in the physical sciences than was his brother since he had served as a military meteorologist. Not only did Tom Odum contribute his own interpretations of these modern ecological concepts to Eugene Odum, he also read and commented on the manuscript of *Fundamentals* and wrote the chapter on energy in ecological systems (Taylor, 1988, 224). It is likely that Tom's interest in biogeochemistry, stimulated by Hutchinson, led Eugene Odum to emphasize biogeochemical cycles along with the ecosystem concept in his textbook.[8]

The environment in which these ideas were presented had changed fundamentally from that existing before the second world war. Not only did the war wipe away many individuals, structures, and institutions, it also created a

4.2 Howard Tom Odum, 1980s. Photograph courtesy of H. T. Odum

reaction to the past that could be characterized as an enthusiasm for reconstruction, growth, and the productive forces to effect change. In America at least, the universities were crowded with returning war veterans. The faculties of universities and colleges expanded, and new opportunities for the funding of ecological research came, first through the Office of Naval Research, and then through the Atomic Energy Commission and the National Science Foundation. In contrast, a preoccupation with the problems of reconstruction prevented the European countries from being as actively involved in ecosystem research as might have been expected from their activity before the war.[9]

In addition to this environment of change, there was another important element recognized by Odum in his chapter on the ecosystem. This was the growing realization that humans were destroying the environment. The story, of course, was not new. In 1864, George Perkins Marsh had described soil erosion in modern terms. William Vogt wrote *Road to Survival* in 1948, and Fairfield Osborn wrote *Our Plundered Planet* in 1948 followed by *Limits of the Earth* in 1953. Rachel Carson's *Silent Spring,* published in 1962, ushered in the environmental era. Her book became a dominant force for environmental action internationally.

These expressions of global and local concern about environmental deterioration engendered a public demand for action that led to a variety of institutional mechanisms to study and manage the environment. Ecology was one of the sciences called upon to provide a scientific basis for environmental management. Ecosystem ecology provided an integrated way to view the environment and therefore was quickly adopted by the public. Ecosystem scientists dealt directly not only with lakes, forests, and fields, but human beings were included as ecological components of ecosystems as well. This was explicit in Odum's book and widely accepted among his colleagues. Ecosystem ecology also was framed, in part, in the languages of engineering and economics and in the new subjects of cybernetics and information. The subject was accessible and made sense to the educated layperson, yet its language and concepts reflected some of the most advanced trends of the 1950s.[10]

Eugene Odum's text fit well into this pattern, and it created a generation of ecosystem ecologists—as distinct from plant ecologists and animal ecologists— who were prepared mentally and technically to contribute to the environmental decades. His book was reprinted several times, revised twice, and translated into numerous foreign languages. As *Fundamentals* was revised, it became larger,[11] whereupon Odum wrote a smaller text, *Ecology,* which was equally successful. Because for a number of years he had little competition in the textbook field, his influence was enormous.

One of the features of the Odum texts was that not only did he synthesize a personal view of ecology from countless monographs and books of others, he also showcased up-to-date ecological work. Odum liberally used data and figures from current publications and cited active ecological studies. As editions of the works appeared, he faced a reduction of the growing literature into his ecosystem format. In the first edition of *Fundamentals* this was less of a problem because ecosystem studies and those studies providing background or foundational information were relatively few. This situation changed rapidly.

STUDYING THE ECOSYSTEM

Studies of whole ecosystems, following the lead of Lindeman, lagged until the first postwar graduate students left university. Among these was Howard T. Odum, who graduated from Yale in 1951 and took a position at the University of Florida where he began studying the springs of Florida, including Silver Springs (Odum, 1957). Silver Springs is a relatively well-defined system, with a single large input (half of the volume comes from a single, large springhead) and a single large output down the Silver River. Even more important for Odum's purposes, the spring is thermostatic and chemostatic, and the volume varies relatively little annually. It thus serves as a natural, constant temperature chamber. Odum called this situation a *steady state* and emphasized that it was a necessary precondition to ecosystem study. Odum pioneered a method of studying system dynamics by measuring the chemistry of the input and output water. The difference between input and output, under steady state conditions, was a measure of the metabolism of the whole system.[12] Tom Odum (1957, 56) was motivated to study the whole system as a unit. His general plan "was to characterize the chemostatic flow, to establish the qualitative and quantitative community structure, to measure the production rates, and to study the mechanisms by which the community metabolism is self-regulated." Here Odum referred to a chemostat, which is a device wherein bacteria may be grown; a constant flow of chemical nutrients maintains the bacteria population. He was drawing an analogy between a chemostat and a spring. Yet the last phrase, referring to the self-regulation of community metabolism, is most significant. Clearly, Odum was thinking of the spring as an ecosystem, with a metabolism regulated by the system itself. This form of thinking paralleled that of Lotka and Thienemann, but Howard Odum does not credit these others for his analogies.[13] His approach seems independent, and while it was not different in kind, it was unique in its intensity and focus.

Springs seemed to be a popular ecosystem for study. At about the same time that Howard Odum was studying Silver Springs, John Teal began a study of a small, cold water spring, Root Spring, in Connecticut, for his doctoral thesis at Harvard University. Teal employed the Lindeman approach of enumerating populations and aggregating population data to describe the dynamic behavior of the whole community. Later, in 1962, after Teal graduated and joined the University of Georgia Sapelo Marine Laboratory, he pulled together data from several sources on the coastal salt marsh into another ecosystem description. Lawrence Tilly, in 1968, reported on the structure and dynamics of another small spring in Iowa, Cone Spring.

During this period limnologists continued studying lakes as systems. For example, in 1953, C. F. Dineen published his doctoral work on a small pond in Minnesota, copying almost exactly the approach of Lindeman, stimulated in part by Samuel Eddy, who also had advised Lindeman to study Cedar Bog Lake. In the marine area, Gordon Riley, Hutchinson's first doctoral student in 1937, and George Clarke, of Harvard University and Woods Hole, were applying these concepts to the ocean off New York and New England. In terrestrial systems where, as in marine environments, it is often difficult to

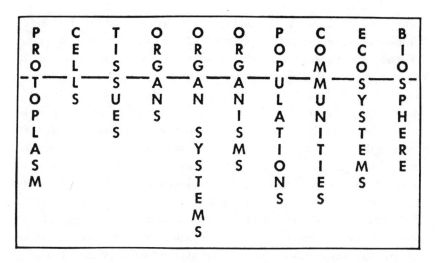

4.3 Biological spectrum of Eugene P. Odum, in which he placed ecosystems within a hierarchy of biological systems. This diagram appeared in his second edition of *Fundamentals of Ecology* (Odum, 1959, 6). Published with permission of W. B. Saunders and Company

observe ecosystem boundaries, ecologists began to apply the ecosystem concept to parts of terrestrial ecosystems, such as the soil. An article from one of these studies (Fenton, 1947) was among the first to use the word *ecosystem* in the title. L. Charles Birch and D. P. Clark of Sydney, Australia, applied the Lindeman approach to soil in 1953, as did biologist Nelson Hairston in 1959 and Manfred Engelmann in 1961. These latter two studies were done on the George Reserve of the University of Michigan, which later became a center for an ecosystem study under the direction of Francis Evans.

Using the word *ecosystem* to refer to the soil component of the community ecosystem indicates a semantic confusion that was characteristic of the time. Evans (1956) in a short note to *Science* magazine, presented the ecosystem as the basic unit in ecology, applicable to the individual, the population, or the entire biosphere. Evans was focusing on the fact that the ecosystem, or ecological system, referred to a system consisting of a biological element interacting with the environment. Evans's position was that the biological unit might be any level of a hierarchy from a cell to the biosphere. Eugene Odum in the first edition of *Fundamentals* took this same position,[14] but in his second edition he used *ecosystem* in a different way. He presented a sequence of ecological units: individual, population, community, and ecosystem (fig. 4.3). In the figure, the ecosystem appears to be equivalent to biological units such as the population and the community. Yet earlier (in 1953) Odum had defined the ecosystem as being made up of the biotic community and its environment. This earlier definition became the general definition of ecosystem for many ecologists educated with *Fundamentals*; others followed Evans and used ecosystem to refer to any ecological system made up of a living part and an environment. Depending upon how the term was used, *ecosystem* could be a new way of saying *ecology*, a way of emphasizing the energy and material transfers underlying all relationships between organisms and between organisms and their environment, or the study of a particular object, the ecosystem. No one was unduly disturbed by these contradictions. Eugene Odum himself used the word in all of its possible meanings. It was only in the later 1960s, when ecosystem studies involved institutions and sources of support, that definitions became important.

As a consequence of the multiple usages, many biological ecologists found that traditional ecological studies on the structure and function of populations and communities qualified as ecosystem studies in the new jargon. Eventually, these became known as "process studies," that is, studies of processes within ecosystems. As we shall see, relatively few ecologists adopted Eugene Odum's concept of the ecosystem as a whole system. Most studies essentially repeated Lindeman's study of Cedar Bog Lake, in that they aggregated the diverse fauna and flora into a small set of trophic groups, determined the flows of energy and materials between groups, and calculated ratios of input to output.

In this period of growth and transformation, ecosystem studies received an important stimulus from an unexpected source, the U.S. Atomic Energy Commission. Atomic energy burst onto the world's consciousness when atomic bombs were dropped on the Japanese cities of Hiroshima and Nagasaki. Development of atomic weapons was pursued with vigor by the United States. In 1945 a test of an explosive device, placed on top of a tower in the desert near Trinity, New Mexico, was one of the first of a long series of nuclear explosions in the environment. Two years later a group of biologists, led by faculty members from the University Of California at Los Angeles, studied the distribution of radioactive fallout in vertebrates, insects, vegetation, and soils at the Trinity site (Larson, 1963). This was the beginning of what became an exceptionally active ecological research area, *radiation ecology*. At the same time, studies of the consequences resulting from the loss of radioactive material from the Hanford, Washington, reactors to the Columbia River were beginning (Fontaine, 1960). From 1946 to 1948 the U.S. government tested atomic explosives at Bikini and Eniwetok atolls in the Pacific ocean. These tests were accompanied by studies of marine ecology.

Philip Gustafson (1966, 67) described the opportunity these radioactive releases provided for ecologists:

No one would consider the deliberate release of radioactivity by weapons tests on a global scale as a means of undertaking an environmental radiation research program. However, such releases have taken place as a consequence of the tests and wide distribution of fission products has resulted. This has made serious study of their effect on the environment mandatory in order to predict and identify the hazards to health. Dramatic opportunities now exist for these studies. In meteorology, with the entire atmosphere tagged with radioactivity, transport and mixing phenomena have been investigated on a global scale. An analogous situation exists in oceanography, although, thus far it has been investigated only to a limited degree. And, fallout has permitted other aspects of "the natural environment" to be investigated by novel and creative means, often along new and unique lines. Fission product debris is only one of many forms of contamination, or pollution, added by man to his environment. The study of radioactive pollution, made relatively easy by today's techniques, should provide an effective insight into the behavior of other forms of pollution such as hazardous chemicals that include carcinogens and pesticides.

At this time, before the rapid escalation of the Vietnam War, many American ecologists were unconcerned that their studies were closely linked to military activities. Rather, they tended to accept the cold war as a fact of life and welcomed the opportunities military research made available. The initial studies described the accumulation of radioactive material in organisms collected at different distances from the source of contamination and at different times after the detonation. Soon it became clear that these scientists were dealing with two different types of problems. First, they were concerned with the physical-chemical problems of radioactive decay and the deposition, absorption, and accumulation of radioactive materials. Second, they were concerned with food chain dynamics and the problems of metabolic turnover, the efficiency of trophic transfer, biological uptake, release, and other problems faced in ecosystem studies.[15] Thus, there was a juxtaposition between the development of ecosystem studies as a part of basic ecology and the practical need to understand the transfer of radioactive materials among ecosystem compartments. Theory and the availability of funds together produced vigorous research activity.

By the mid-1950s, full-scale research programs were organized by the AEC in the United States at its production and test facilities. In 1952 Eugene Odum received one of the first contracts to do research at the Savannah River Plant in South Carolina. Here, with his students and associates, he began studies of old-field succession on the 300 square miles of abandoned land around the reactors. Lauren Donaldson was investigating radioactivity in fish in the Columbia River, and Kermit Larson headed a research team at the Nevada test site. In 1954, Stanley Auerbach began the ecological research program at Oak Ridge National Laboratory in Tennessee, and Jared Davis began a similar program at Hanford (Wolfe, 1967). By 1955, John Wolfe, formerly an Ohio State University professor of botany, became part of the Division of Biology and Medicine of the AEC in Washington, bringing ecological expertise to the highest levels of the agency for the first time. Wolfe had a two-year appointment, which is common for American academics working with the government, but he was so successful that in 1958 he was offered and accepted a full-time position as chief of the Environmental Sciences Branch (Sprugel, 1975). Wolfe was given funds to support research programs, some of which were rooted in ecosystem theory. In 1955 the first international conference on the peaceful uses of atomic energy was held in Geneva. This conference stimulated further research activity worldwide.

Thus, by the 1960s active research programs on ecosystems were under way in the United States, supported mainly by the AEC. These included studies of warm deserts at the Nevada test site led by Larson, Fred Turner, Norman French, and Bill Martin; cold deserts and the Columbia River at Hanford, headed by Davis; sandy coastal plain in South Carolina at the Savannah River plant, headed by Eugene Odum and me; and deciduous forests both at Oak Ridge National Laboratory, headed by Auerbach, and the Brookhaven National Laboratory, New York, headed by George Woodwell. In addition to these studies, there was an examination of the possible use of nuclear explosives for excavation in Alaska (Project Chariot), Panama, and Columbia. Also, there were studies of the radiation effects on the tropical rain forest at El Verde, Puerto Rico, under Howard Odum, as well as studies of coral reefs and ocean ecology at Eniwetok atoll.

Wolfe and the AEC had created an invisible college, in the sense used by Crane (1972). This community exchanged members, met as a group in national symposia in 1961, 1965, and 1967, and had a joint research agenda. Their major objectives were: (1) predicting the movements of radioactive material through ecosystems and the impact of radiation on ecological systems, including whole organisms; and (2) the application of radioecology techniques to solving problems in basic ecology (Odum, 1965). Of course, many individuals outside of the research centers were funded by the AEC. There were also active support programs for ecology in the Office of Naval Research, the National Science Foundation, and a few other government agencies. The teams, however, were centered at the AEC national laboratories.

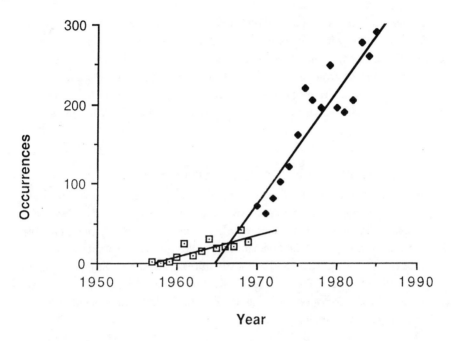

4.4 The frequency of the term *ecosystem* as it appeared in published titles and key words of articles, as abstracted by *Biological Abstracts*. (The squares with dots indicate pre-1969 articles, and the solid squares, post-1969 articles.) Data furnished by Biosis, Philadelphia.

Scientists from other countries did not experience quite the same juxtaposition of interest and support as did U.S. ecologists. Not only were national funds and aid required for reconstruction, but military programs were proscribed in Japan and Germany. Further, in Germany there was active hostility toward holistic thinking, which had provided a scientific base for national socialism. Nevertheless, the interest in ecosystems was increasing throughout the world, especially in studies of the dynamics of ecosystem compartments, in biological productivity, and in the transfer between elements in food chains. In 1957 the word *ecosystem* was first tabulated in the indexing system of *Biological Abstracts*, and the appearance of the word in the title of articles began a continuous increase (fig. 4.4). The break in the curve in 1969 indicates the time when the next major event in ecosystem studies occurred, the International Biological Program.

Thus, by the mid-1960s the ecosystem concept had become an organizing idea in American, if not international, ecology. One may conclude that the ecosystem had become a scientific paradigm, in the sense used by Thomas Kuhn (1962), with a community of investigators interacting in pursuit of roughly defined, common research goals, and with a set of nascent institutions where teams of ecologists were funded to carry out long-term studies of the structure and function of ecosystems. As research was carried out in the formative period, the nature and limitations of the ecosystem concept became clearer.

IMPLEMENTATION OF THE ECOSYSTEM CONCEPT

In the post–Lindeman era ecosystem studies progressed in two directions. There were studies that attempted to emulate Lindeman and study a system as a whole. These were the least common, probably because they required intense work over long periods and demanded a philosophical acceptance of the concept of the whole system. Studies of structures or functions within ecosystems were more frequent. These studies focused on a few properties of the systems, such as food webs, food chains, trophic levels, productivity, metabolism, energy flow, and ecological succession.

Ecosystem Structure

The first problem faced by an ecologist attempting to study a whole system is counting the living organisms that make up the system. There is nothing new about this task. Ecologists have been enumerating the biota from the beginning of ecology. Complete counts of limited areas, such as the census of four square feet of soil by McAtee (1907), revealed that the fauna and flora were composed of hundreds and sometimes thousands of species. In many cases ecologists collected species new to science in these censuses. Further, they found that the species in a biotic community were not constant, as Lindeman had observed in Cedar Bog Lake. Thus, ecologists found it necessary to reason from censuses of species in defined samples of the community to the patterns of species abundance of entire communities.

Careful counts of the species showed that they were organized into predictable structures (Gleason, 1922; Williams 1944, 1953; Preston, 1948, 1960, 1962). Often a few species were abundant; the majority were less abundant or rare. Graphing the cumulative species observed against the number of individuals counted—when converted to logarithms—frequently produced a straight line relationship. Several investigators pointed out that the slope of this line was a measure of the diversity of the community (Fisher, Corbett, and Williams, 1943).

These studies, while extremely important to ecology as a whole, were only indirectly relevant to ecosystem studies. The problem was not how many species there were but how the species used food and energy. Lindeman solved this problem by focusing on the dominant species, determining their food habits, feeding coefficients, metabolism, and growth, then extrapolating from the dominant to the rare species under the assumption that the rare species acted similar to the dominants. There was little evidence that this assumption was correct. Since the focus was the food cycle, the assumption meant that organisms feeding on plant material consumed and digested food at rates more like each other than at rates like those of carnivores or other types of feeders. Though this may seem a logical conclusion, it is actually a process of circular reasoning. Trophic levels were considered to be subsystems of the whole system, and therefore by definition all members of the subsystem shared the same properties. After an organism was placed in the correct subsystem, it could be treated mechanically without further reference to its biological character. Taxonomically related organisms were members of the same subsystem, unless there was evidence that suggested otherwise. Lindeman led the way in applying this method and was soon followed by others, such as Howard Odum in his studies of Silver Springs (fig. 4.5). Indeed, this model was so universal that Eugene Odum (1968) referred to it as an "Odum-device."

It was not until 1971 that a trophic model different from that employed by Lindeman was proposed by experienced ecosystem scientists Richard Wiegert and Dennis Owen. The prime motivation for altering this familiar model was an argument about controls within systems. They argued that the mechanism limiting any heterotrophic population, as contrasted to a heterotrophic level, depends on whether the organisms using the resource directly affect the rate of resource supply and on their life history characteristics. They also felt that Lindeman's model was inadequate to represent the dynamic behavior of ecosystems since the consumer food chains were defined as a cascade and all the unused energy from the producers and the consumers was channeled into a catchall compartment called decomposers. The Wiegert and Owen model is shown in figure 4.6. They defined three major groups: *autotrophs*, which replaced Thienemann's producers; *biophages*, which replaced herbivores, carnivores, and top carnivores; and *saprophages*, which replaced decomposers. Autotrophs are organisms that use energy sources other than the chemical potential energy of organic matter to synthesize organic compounds from inorganic elements and compounds. Biophages are organisms that use living organic material for food. Saprophages are organisms that use nonliving organic materials for food. The biophages and saprophages may be subdivided in order to account for energy transfer chains of any length. This form of organization does have advantages over the Lindeman model because it explicitly

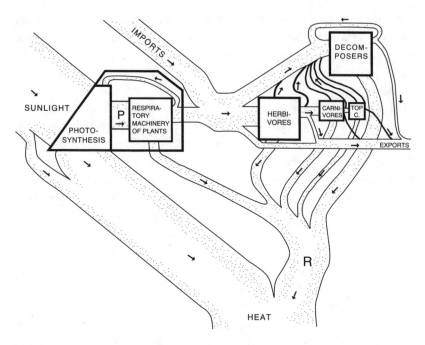

4.5 Howard T. Odum's diagram of energy flow in natural communities, used for his study of Silver Springs, Florida (Odum, 1957). In the diagram, *P* symbolizes the gross primary production, *R* is total community respiration. Trophic levels are indicated by square boxes, and the size of flows are roughly indicated by the width of the stream. Published with permission of the Ecological Society of America

shows the linkages between biophage populations and saprophage populations at various distances from the autotrophs, but it has not been widely adopted.[16]

Another alternative model was suggested by Howard Odum, John Cantlon, a plant ecologist at Michigan State University, and Louis Kornicker, a colleague of Odum in Texas. When discussing their joint work at an ecological meeting, these three recognized that the well-known straight line relation between cumulative species and the logarithm of individuals was the same formulation that Ramon Margalef (1957), a Catalonian ecologist interested in information and the species diversity of plankton, had used to describe the information content in ecological communities. Odum, Cantlon, and Kornicker (1960) reasoned that ecological communities might be organized like human communities, with the species representing different occupations in nature having different information contents. The larger the human community, the greater the number of occupations that can supported by the population. Rare occupations, such as an optometrist, are not found in small

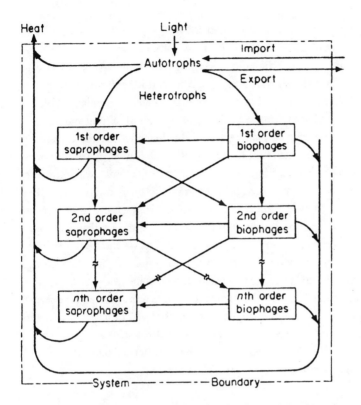

4.6 Model of trophic transfer proposed by Wiegert and Owen (1971). The model distinguishes between living and dead organic material, permits an open number of steps in the transfer to be represented, and stresses the interaction between levels. Published with permission of the *Journal of Theoretical Biology* and Academic Press

communities. This analogy was not original with these three as it had been used earlier by Charles Elton in his definition of the niche. Elton (1927) suggested that the niche of an organism was like the profession of a person. Thus, if species are like occupations, then it would follow that a graph of species and individuals could be considered a representation of a hierarchical organization of the ecosystem. Deviation from a straight line would suggest a suborganization of the system, sampling problems, or fuzzy boundaries between ecosystems. This idea, which brought together concepts of species diversity and ecosystem organization, was never developed further.[17]

Thus, ecologists studying ecosystem structure chose the Eltonian pyramid model over the Eltonian food web model. Ivlev's (1945) cogent criticism that the trophic level approach did not include the real feeding habits of species was ignored, or at least was never discussed until the idea was expanded by Hairston, Fred Smith, and Lawrence Slobodkin (1960)—all biology professors at the University of Michigan. In retrospect, the problem appears to have involved more than simply the difficulty of dealing with a myriad species. It was exceptionally difficult to census all of the species and identify them properly. Taxonomy still has not advanced to a point where all species are identifiable by name. Also, while the study of food habits was active and of long standing, it was inadequate to the task of placing each individual encountered during an enumeration of the fauna into the proper trophic level. Rather, it was necessary to reason by analogy, so that if one member of an order or family was known to eat plant food, then another member also must be an herbivore. And the enthusiasm present at the time for a physical or engineering approach to systems tended to deemphasize the significance of biological differences. Species and individuals were represented as mass, energy, or chemical elements, and their biological reality disappeared except to define the links of feeding. Organisms were expected to act mechanically in predictable ways, and from this perspective, it seemed reasonable to expect that there was a limited set of ways to act. In other words, it was thought that the observed differences in detail were not as important to the overall behavior of the system as the common features of the behavior of the individuals. This abstraction allowed a level of synthesis that would not have been possible otherwise.

The Dynamics of Ecosystems

The ecosystem approach provided the basis for a functional view of natural systems. This was its key element. The functional or dynamic aspect basically had two parts. One part was concerned with the energetics of systems, and the other was involved with organic productivity. Although each is part of the other, the methodology and application used for studying each resulted in different developments. Ecological energetics quickly became a subject of its own, with many studies undertaken on the energetics of species populations and individual organisms. It is difficult to separate the part of ecological energetics that is relevant to an understanding of ecosystems from the part that is applicable to other topics. (See Phillipson, 1966; Odum, 1968; Golley, 1972; Wiegert, 1976.)

Energy Flow

The energy flow approach was central to Lindeman's trophic-dynamic aspect of ecology, although he developed the fuller implications of the energy approach only after he came into contact with Hutchinson, Deevey, and others at Yale.[18] Even though Lindeman recognized that energy permitted an integration of diverse activities, he actually used energy as a currency and not as a fundamental theory of ecology. Juday (1940), also used energy as a currency. Even Lotka (1925), who explicitly applied thermodynamic theory to ecological systems, did not use thermodynamics directly except in his explanation of evolution, which he interpreted as the second law acting within biological systems.

This situation changed decisively with the appearance of Eugene Odum's ecology textbook in 1953. In chapter 4, "Principles and Concepts Pertaining to Energy in Ecological Systems," he reviewed the fundamental concepts related to energy,[19] stating: "Energy is defined as the ability to do work. The behavior of energy is described by the following laws. The *first law of thermodynamics* states that energy may be transformed from one type into another but is never created or destroyed. . . . The *second law of thermodynamics* may be stated in several ways, including the following: No process involving an energy transformation will spontaneously occur unless there is a degradation of the energy from a concentrated form into a dispersed form." This could not be clearer! Odum then proceeded to apply the energy theory to the food chain, pyramids of numbers, biomass, and productivity. The generalizations Odum developed from these concepts almost all relate to the ratios of energy or biomass in trophic levels. The ideas that larger organisms appear higher in the food chain, energy is lost at each transfer up the chain, and so forth, were already present or anticipated in Lindeman's articles.[20]

Thus, the energy flow approach that entered ecology as a way for comparing diverse biological processes was transformed into a theoretical foundation for ecosystem function. Although most ecologists were inadequately prepared in the physical sciences to take advantage of the energy concepts and language—except in a general way—adoption of the energy approach provided two useful tools for ecological analysis.[21] First, by following the first law of thermodynamics, ecologists studying the energy flows in a defined system could expect that all the energy entering the system would equal all the energy leaving the system, plus that stored within it. Thus, the first law provided a balance sheet that could be used to check the accuracy of the measurements of inputs, outputs, and storages. Second, according to the second law, at each transfer of energy in the food cycle, energy is lost to a heat sink and is no longer available to do further work. Therefore, the efficiencies of energy transfer must be less than 100 percent.

Ecologists began to apply these ideas to all levels of ecological organization. Relatively few could or would tackle the whole community. Most focused on a single trophic level, such as vegetation, or producers, a food chain, or a single species or taxon. A few developed laboratory systems. These studies began in several places at about the same time. Howard Odum was active first in Florida and later at the University of Texas marine laboratory at Port Aransas. Slobodkin and his students extended the work from his doctoral thesis on the biology of laboratory colonies of Daphnia to the energetics of Daphnia and other small aquatic organisms at the University of Michigan. Teal, who began his research with a Harvard doctoral dissertation on Root Spring, moved to the University of Georgia and together with Eugene Odum and his students applied concepts of energetics to the salt marsh. Odum also was using energetics in his studies of old-field succession at the AEC Savannah River plant. I had carried out a study of the energetics of a food chain for my doctoral thesis with Don W. Hayne at Michigan State University and joined Odum in these old-field studies in 1958. Wiegert, who was also associated with Hayne for his master's thesis and with Slobodkin when he was a doctorate student of Evans at the University of Michigan, joined the Eugene Odum team in 1962. Eugene Odum and Howard Odum together studied the dynamic behavior of a coral reef on Eniwetok atoll in 1954. These early applications of energy concepts to ecological systems were carried out by a relatively small group of ecologists. They were all exploring the implications of the laws of thermodynamics when used in these new settings.

The initial studies were almost entirely descriptive because essentially nothing was known about the energetics of ecosystems. It was necessary to construct a body of observations so that the patterns of behavior could be defined and then explained. The hypotheses that structured the descriptive studies, however, tended to be declaratory statements derived from thermodynamics or the operation of a machine. Lotka had set the stage for this approach, and Eugene Odum's textbook format firmly established it. Lotka reasoned analogically from physical theory to biology. In his introductory comments (Lotka, 1925, 325) on energy kinetics, for example, he explains: "These components—aggregates of living organisms—are, in their physical relations, *energy transformers*. . . . The dynamics which we must develop is the dynamics of a system of energy transformers, or *engines*." He continued with an explanation based on the steam engine. Lotka's capacity to find analogies was stimulating, but it required the biologist to accept the analogy between organisms and heat engines. Not everyone could do this. In Odum's text, these analogies were not used. There, the biologist had to accept the theory that biological systems operate according to the physical laws of thermodynamics—an easier leap to make for most of us. It is essential for understanding this developmental phase of ecosystem studies to recognize that it began in analogical thought, authoritarian statements, and physical laws—not in questions.[22]

The first issue in ecosystem energetics concerned the balance of the whole system, as required by the first law of thermodynamics. Did the inputs of energy to an ecosystem or component of an ecosystem equal the outputs? While many ecologists assumed that the first law held true and used it as a tool to determine if their measurements were correct (for example, Teal, 1957, 1962), Howard Odum (1957) had a situation where he could measure directly the system's production and metabolism under steady state conditions. In Silver Springs, Odum found a discrepancy between income (6,510 g/m²/yr) and losses (6,766 g/m²/yr) of 256 g. His data are such that we cannot tell whether a 4 percent difference is significant or not. Odum's own conclusion is ambiguous: "Although an exact agreement between production and respiration-organic matter loss is not found, the difference of 4% can presumably serve as some estimate of the reliability of the orders of magnitude. The estimates of respiration and organic export are certainly rough approximations, whereas the production measurements may be more accurate" (Odum, 1957, 106).

The balance between energy input and output was expressed by the ratio of production to respiration, or the P-R ratio. Odum began a study of the P-R ratios in a variety of waters. He and Eugene Odum discovered another 4 percent difference between input and output on Japtan Reef in July at Eniwetok. In a theoretical review (H. T. Odum, 1956a) he compared P-R ratios for a variety of aquatic systems and showed that there was a considerable range of values. Clearly, not all systems are in balance. Odum erected a classification of water bodies based on their P-R ratios, which separated autotrophic from heterotrophic ecosystems. By autotrophic Odum meant ecosystems obtaining their energy by the photosynthesis of their own producers. Heterotrophic systems, such as caves, require an organic energy input from outside the system. The most thorough test, however, was made by Odum's student, Robert Beyers (1963), who showed that in laboratory microcosms under steady state conditions and after a period of successional development, the P-R ratio averaged 1.05, based on 120 determinations. The departure of the average ratio from unity in these diurnal curves is only 0.09. Beyers commented that "the departures of these ratios from unity is well within the limits of experimental error of the method." His conclusion that the ratios showed the microeco-systems were truly balanced is strong evidence for the question concerning balance in the input and output of energy in ecosystems.

A second question introduced by this approach concerns the relative significance of the rate of energy transfer across a food link and the efficiency of that transfer. Lindeman, influenced by Hutchinson, discovered that consumers higher in the trophic pyramid are more efficient in using their food than organisms lower in the pyramid. These efficiencies provided a simple way to compare different parts of systems, such as comparing herbivores to carnivores. They were also the beginning of a mathematical treatment of systems (Clarke, Edmondson, and Ricker, 1946). Most ecologists interested in the analysis of systems energetics calculated efficiency ratios. Howard Odum used six different ratios in his analysis of Silver Springs. Yet the attempt to go beyond the mere reporting of ratios toward theories of energy transfer resulted in the first serious disagreement in systems energetics.

Hairston, Smith, and Slobodkin presented a brief article, published in *The American Naturalist* in 1960, on the control or regulation of populations organized in different trophic levels. They observed that fossil fuels accumulate at a negligible rate compared with the rate of photosynthesis. If all the energy that is fixed in photosynthesis flows through the biosphere, they reasoned that the organisms taken together must be limited by the amount of energy that is fixed. They pointed out that this must be especially true for the decomposers (or saprophage food chains), otherwise we would observe an accumulation of organic matter and fossil fuel. They also observed that terrestrial plants were resource limited, the resource usually being light, but it also can be water, climate, or mineral elements. Since herbivores seldom eat all the plant material available, it must be true that herbivores are not limited by their food. Carnivores, however, usually control the number of herbivores and are food-limited. The authors commented: "There thus exists either direct proof or a great preponderance of factual evidence that in terrestrial communities decomposers, producers, and predators, as whole trophic levels, are resource-limited in the classical density-dependent fashion" (Hairston, Smith, and Slobodkin, 1960, 423). The consequences of these observations are: (1) those communities will be most persistent where herbivores are controlled by carnivores and producers are resource-limited, and (2) interspecific competition for resources exists between producers, carnivores, and decomposers. Herbivores may be exceptions to density-dependent regulation.

In their assertion, Hairston, Smith, and Slobodkin were thinking analogically from the density regulation of populations to the regulation of trophic levels. If a population was density dependent, it meant that the population was regulated by factors that came into effect at a specific density. The alternative was density-independent regulation where density was regulated by factors that were independent of numbers, such as the weather. The relative importance of density-dependent and independent regulation was an ongoing debate among population ecologists, especially after H. G. Andrewartha and Birch (1954) in Australia put forward their strong advocacy for density-independent control.

Hairston, Smith, and Slobodkin were applying population regulation concepts in a reasonable way, but their choice of the density regulation model attracted the attention and criticism of population ecologists (Murdoch, 1966; Ehrlich and Birch, 1967; Wiegert and Owen, 1971). These criticisms were mainly of definitions or philosophy. William Murdoch as well as Paul Ehrlich and Birch, as population ecologists, assumed the authors were discussing the regulation of populations and denied the existence of trophic levels. Indeed, according to Murdoch (1966, 223): "In fact, organisms exist in populations and it is doubtful that nature is organized around the trophic level with regard to the processes involved in the limitation (of) numbers." Ehrlich and Birch (1967, 103) had a similar perspective: "If we can draw any general conclusion from the work which has been done on natural populations, it is that single, neat 'control' mechanisms are unlikely to explain fluctuations in the size of single populations, let alone numbers of all organisms of a trophic level." Hairston, Smith, and Slobodkin made it clear that they considered it legitimate to argue logically from trophic level to population. Yet this procedure is not valid. In the first place, a "trophic level" exists only as an abstraction.

Murdoch (1966, 224) also raised questions about the form of ecological hypotheses. His comment was: "The particular instance of the non-testable idea which has been criticized here (if it is such an instance) is not unique in ecology, and such ideas seem to arise from attempts to reach statements about a broad range of undefined phenomena before statements about recognizable classes of events are achieved." Murdoch was calling for the establishment of operationally defined, testable hypotheses and was pointing out a weakness in ecosystem study. As far as I know, this was the first published criticism of a type that was to become increasingly common. This was the Achilles' heel of ecosystem work, and Murdoch's comment was a signal that observers outside ecosystem studies saw the organizing, pattern-recognizing phase of ecosystem science as over and felt that the studies should become increasingly cast in conventional scientific terms and operations.

Thus, in the mid-1960s we find a continuation of some of the same arguments Tansley and Thienemann were dealing with thirty years earlier. Some ecologists felt that all ecological phenomena must be interpreted at the level of the individual organism, organized into populations. This view had already driven the phytosociologists away from ecology to form the field of vegetation science and had isolated some applied ecologists from basic ecology. Now the holders of this view were seeking to marginalize those ecologists interested in systems of large spatial scale from what they considered the central field of ecology. Slobodkin, Smith, and Hairston (1967), in their response to this criticism, emphasized that they were discussing trophic levels and not single populations, clarified and expanded their definitions and examples, and cited newly published data that confirmed their predictions. They declared themselves in fundamental disagreement with their critics.

Yet one criticism came from Wiegert and Owen. As former students at the University of Michigan, they were familiar with most of the participants in the debate, including Bill Murdoch. Their comments proved useful in revising the Hairston, Smith, and Slobodkin concept. Wiegert and Owen emphasized that the argument depended not so much on the presence of unconsumed resources, but rather on the potential impact of the organisms on the supply of resources. For example, saprophage (decomposer) populations have little ability to control directly the activity of autotrophs, just as autotrophs have little ability to control the amount of solar energy available. This observation led them to distinguish terrestrial systems dominated by autotrophs of large size from aquatic systems dominated by autotrophs of small size. The two kinds of systems perform differently and have different biophage networks. The terrestrial systems typically have three-link chains of autotrophs, first- and second-order biophages, high-standing crops of autotrophs, the low density of first-order biophages, and a low percentage of primary production used by the biophages directly. In contrast, the aquatic systems tend to have four levels, with the addition of third-order biophages in the chains, low-standing crops of autotrophs but high production and turnover, moderately high density of first-order biophages, and little of the primary production going to the saprophage food chains. Thus, Wiegert and Owen, accepting the assumption of trophic-level structure, shifted the attention from population regulation to the control of resources—from a focus on the organisms in one level to the interactions between two levels. This proved to be a useful clarification.

Howard Odum and R. C. Pinkerton (1955), a physicist at the University of Florida, also addressed these questions from a theoretical perspective. Their proposition was that natural systems tend to operate at that efficiency that produces the maximum power output, not the highest possible efficiency: "Under the appropriate conditions, maximum power output is the criterion for the survival of many kinds of systems, both living and non-living. In other words, we are taking 'survival of the fittest' to mean persistence of those forms which can command the greatest useful energy per unit time (power output)." Their argument was developed using a physical analogy in which the transfer of energy involves two parts: (1) a release of stored energy, a decrease in free energy, and a creation of entropy, and (2) the storage of energy, an increase in free energy, and a decrease in entropy. The coupling of input and output produces the different rates and efficiencies observed. Odum and Pinkerton examined these couplings in a range of different open systems, from a machine to a climax community to a civilization. They concluded that all the systems performed at optimal efficiency for maximum power output and that this optimum efficiency is always less than 50 percent. In another article a year later (Odum, 1956a), Odum—exploring data on the efficiency and production of natural plankton communities in Trout Lake, Wisconsin, collected by Man-

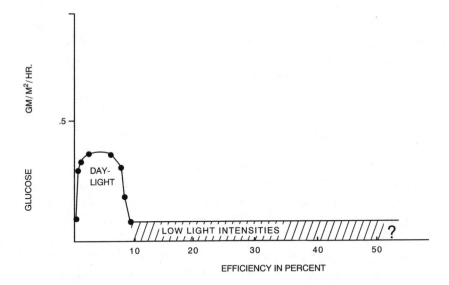

4.7 Odum and Pinkerton's (1955) illustration of the maximum power output theorem in primary production. Glucose production rates, produced through photosynthesis, are maximum at low efficiency. Efficiency is calculated as the energy output in photosynthesis divided by available sunlight. The data were originally from Manning, Juday, and Wolf (1938), who measured photosynthesis in lake water using the light-dark bottle method. Published with permission of *The American Scientist*

ning, Juday, and Wolf in 1938—showed that when efficiency is plotted against primary production (fig. 4.7), there is a hump-shaped curve with a long tail toward high efficiencies at low light intensities. In this system, highest production occurred at an efficiency below 10 percent.

Odum reasoned from these observations that competitive communities would tend to develop a similar low efficiency when exposed to similar light regimes and have a similar photosynthetic output per area. Since the metabolism per gram of heterotrophic organisms is inversely related to their body size as a two-thirds power function (Zeuthan, 1953), Odum concluded that communities made up of small plants, such as plankton, should have a smaller biomass to achieve the same output per area as a community made up of large and metabolically less active plants, such as a forest. Teal's (1957) compilation of the ratios of energy stored in plant production to energy received in solar radiation showed that the ratios in all communities were less than 10 percent. Odum and Pinkerton's maximum power output theorem became a central tenet of Howard Odum's energy theory (Odum, 1971).

The metabolism-to-body-size relationship was widely used in the analysis of the energy flow of organisms. These relationships were determined by placing organisms in metabolic chambers and constraining their movements. Since animals are normally active in nature, it was asked whether the real metabolic activity of a free-living organism was greater than or the same as that of an organism confined in a metabolic chamber. This question was addressed by two lines of work. First, ecologists attempted to provide all the necessary resources in the metabolism chamber so that the organism could carry out its life functions. Of course, this approach was more successful for small organisms than for large ones, and Wladyslaw Grodzinski, of Krakow, Poland (Grodzinski and Gorecki, 1967; Grodzinski and Wunder, 1975), was exceptionally successful in partitioning the energy requirements of such processes as feeding, pregnancy, and lactation in his studies of the energetics of mice. The second effort was toward finding a way to monitor the metabolism of free-living organisms. In one attempt, a radio was attached to an animal in order to follow its movements (Lord, Bellrose, and Cochran, 1962; LeFebvre, 1964) and to record its heart rate (Folk and Hedge, 1964). In another attempt, a metabolically active chemical was injected into the animal, and the chemical's turnover rate was followed (LeFebvre, 1962, 1964). Finally, the turnover of radioistopes was used to measure metabolism (Tester, 1963).

The data available before 1965 (Golley, 1967) suggested that the metabolic cost of free-living activity was no more than twice the resting metabolism in most organisms. Of course, at times of high activity, such as with a bird in flight, the rate could be greatly accelerated. Even the mice in Grodzinski's apparatus, however, did not exhibit levels that were more than twice the resting rate at most times in their life cycle.

Finally, a question was raised about the energy stored in the tissues of organisms. These energy contents seemed to vary considerably. Was there a standard value for a specific organism? On the assumption that there were standard values of energy content similar to the standard values of metabolism and chemical content being gathered in the handbooks of the period, I made a comparison of the energy content of naturally occurring materials (Golley, 1961). I wanted to provide a data source for ecologists studying natural communities that could be used to convert data on biomass into energy values without having to go through the laborious process of determining the energy content in a bomb calorimeter. Energy values varied for the kind of organism, the stage of growth, the time of year, and the habitat. A small industry was formed to work out the details of these differences, leading ultimately to a large compilation of energy values (Cummings and Wuycheck, 1971).

At the same time, Slobodkin and Sumner Richman (1961) were considering the caloric values of animals from a statistical perspective. Their hypothesis was that three statistical distributions of energy data were possible: (1) a normal, symmetrical distribution of energy values, which would suggest that all animals shared the same biochemical constitution, (2) a skewed distribution with the mode near the lower range limit of energy content, which would suggest that high caloric content was not selected by evolution—rather excess energy was used for reproduction, and (3) a distribution with several peaks showing significant differences between taxa. The distribution found by Slobodkin and Richman fit the second case;[23] twelve of seventeen determinations fell between 5,400 and 6,100 calories per ash-free gram dry weight. One record for a brachiopod was quite low at 4,397 calories; the highest value was for an insect (a spit bug, *Philenus leucophthalmus*): 6,962 calories.

The pattern proposed by Slobodkin and Richman was quickly challenged. Wiegert (1965) examined the caloric content of all life history stages of the meadow spittlebug, *Philaenus spumarius,* and found that the caloric values for this one animal practically spanned the range of values found for seventeen species of animals studied by Slobodkin and Richman. The lowest caloric value found by Wiegert was for the first instar (4,976 calories per ash-free gram), and the highest was in eggs (6,503 calories per ash-free gram). Both Wiegert in this article and Slobodkin (1962) in a later monograph concluded that the energy content of an organism was a valuable piece of information in ecological research, and other information on the life history and ecology of the species was needed to interpret these data.

Robert Paine of the University of Washington, reporting on the caloric value of marine organisms, pointed out in a detailed explanation (Paine, 1964) that there were several errors possible in caloric determinations. One involved the decomposition of salts, such as $CaCO_3$, at high temperatures in a bomb calorimeter. Also, correction for the water of hydration in salts of many marine species was needed, since the normal drying method did not remove this water. When Paine corrected his data, he found that the values for the animals ranged from 4,943 to 6,675 calories per gram ash-free dry weight. When these records were added to those of Slobodkin and Richman, a symmetrical distribution around an average value of 5,700 calories was produced. Paine stated that a conclusion about the distribution of caloric value in organisms was premature.

The energetics models all focused on the energy captured through photosynthesis and transferred through the food cycle. The energy models formulated by Howard Odum (1957) and others, however, showed that a large proportion of the incident solar energy was not used by the ecological community in the food cycle. In Silver Springs, for example, insolation was computed to be 1.7 million kilocalories per square meter per year. Yet the amount used directly by the plants in photosynthesis was only 20,810 kilocalories. What happened to the energy that was not used in the food cycle? Was it lost in some way and therefore of no further interest to ecologists?

Juday's (1940) examination of the heat budget of Lake Mendota made the answer clear. The bulk of the incoming solar energy created the energy environment of the living organisms. In Lake Mendota it caused the overturn of the lake in spring and fall that was essential to lake function, and it caused the winter ice cover and the summer thermocline. These energy environments, in turn, created the selection force on plants and animals that dictated their roles in the food cycle, their energy contents, and energy flows. Connections such as these, however, were not followed explicitly by those researchers developing the ideas of Lindeman, especially those concerned with the energetics of terrestrial systems. Lindeman focused on the food cycle, and it was not until David Gates entered the picture in 1962, rather late in the era we are considering, that the heat balance aspect of energetics was again treated as a factor of importance.

Gates, a physicist with an interest in the interaction between the atmosphere and plant canopies, became active in ecological research in the early 1960s. He presented his ideas about heat balance and ecological systems first by lectures at the Universities of Minnesota and Colorado and then in a monograph, *Energy Exchange in the Biosphere*. This 151-page volume introduced ecologists to atmospheric physics, solar and thermal radiation, and methods for measuring heat balance. Gates was a frequent participant in ecological meetings at this time, often demonstrating the heat balance of leaves and organisms with his hand-held net radiometer. Gates emphasized the need for measuring the heat balance directly, and he showed how leaf shape and inclination, animal movements, and distributions were governed by their energy environments. He and his student, Warren Porter, developed a model (Porter and Gates, 1969) of the energy environment of an individual organism, which described the dimensions of an animal's energy niche. He thus created a basis for incorporating heat balance equations within biological energy models.

Productivity

The concept of production has been of importance to ecologists, especially applied ecologists, from the beginning of the subject. The yield to man from harvesting a product, the yield taken to feed a predator, or the production of plant material in a shallow lake are obvious quantities to the layperson and ecologist alike. Yet when one begins to measure production and apply the measurement to understanding populations and communities, complications arise. According to Amyan Macfadyen (1949), Thienemann (1931) was one of the first who attempted to assess the definitions of production and productivity.

He distinguished between the quantity of organic matter produced per area, the quantity produced per area per time, and the quantity produced by organisms, including their excreta released during the time of production. All of these quantities are measures of organic production, but each represents a different process. Other ecologists did not always use Thienemann's distinctions. Rather, they formulated concepts of production that fit their own studies and circumstances, producing a chaos of terminology.

Macfadyen reviewed this confusion of terms and attempted to equate the usage of the various authors by distinguishing between the concepts that were related to the amount produced and those related to the amount produced per time; the latter is a rate process. Macfadyen termed the first, *production,* and the second, *productivity.* This usage generally has been followed ever since, but Macfadyen did not clear up all the confusion by this semantic analysis.[24]

The terms may be clarified if we examine the flows of energy within an ecosystem. Energy comes into the system from a source outside of the system; the source may be the sun, it may be organic material as seen in Teal's Root Spring, or it may be the available herbivore flesh to be eaten by a predator population. Once energy enters the system, it may be assimilated. Assimilation implies that not all of the energy entering can be used or absorbed by the system; in other words, it distinguishes between feeding and digestion. This distinction is important for vertebrates and is relatively easy to understand in that context. It is not as well understood for other organisms. The assimilated energy then is used for all of the activities of the organisms, including energy metabolism, reproduction, growth, and excretion. During the process of living, the organism excretes material and also produces wastes. Eventually organisms die. If we assume that the system is in a steady state, so that at the beginning and end of the period of observation the system contains the same amount of energy or material, the product is the organic matter leaving the system, and it may serve as a source to another system, or it may be heat energy that is no longer capable of doing work in the biological system. Productivity is the rate of product formation specified by area or volume.

This simple description is complicated when applied to plants or animals. In plants, solar energy enters the leaf and is fixed in photosynthesis, which is comparable to assimilation in an animal. Some of the photosynthate is used in the plant for its metabolic needs, and the remaining energy or photosynthate is available to other organisms. This latter amount is termed *net primary production* to distinguish it from the amount fixed in photosynthesis, which is called *gross primary production.* Plant production is identified as *primary production,* since it is the first step in the typical food web.

Howard Odum (1956b) was one of the first to note that Lindeman's use of the word *production* deviated from this simple scheme. Lindeman, influenced by Hutchinson, used *productivity* to mean the rate at which energy was taken in by the system. That is, Lindeman considered the positive part of energy change, or the rate at which energy was contributed to a trophic level, as "true productivity." Kozlovsky (1968) observed that true productivity in this sense referred to the gross primary production of the autotrophs and assimilation of the animals. Unfortunately, this usage was repeated by Clarke (1946) and by Allee and co-workers (1949) in their textbooks.

Further confusion was engendered by Thienemann's (1931) assertion that it was impossible to measure productivity by adding up the increments of the growth of organisms over an entire year. Macfadyen pointed out that Thienemann confused the flows of materials with energy. When one considers the ingestion and assimilation of food and the excretion of organic products, the concept of a food cycle is clear. Organic material and the chemical elements in these materials cycle in and out of the trophic levels or populations. Macfadyen emphatically states that "whereas energy passes only once through an ecosystem and is utilized only by one animal on one occasion, matter is continuously either circulating or in store; it does not leave the system (by definition)."[25] Matter, then, can be regarded as a vehicle for energy. We shall see latter that this solution was not acceptable to all ecologists.

Finally, many ecologists confused the meaning of the term *productivity* with that of *standing crop*. Standing crop refers to the amount of material present at a moment in time. It is what is measured directly when a quadrat of grass is harvested or a population of animals is censused. If these observations are repeated over a growing season, an increase in the standing crop is observed. This is due to the production process. The difference between the first and last observation divided by the time is a rough approximation of the crop's productivity. It is rough because in vegetation it does not include the productivity that has been metabolized, eaten by herbivores, died and fallen to the ground as litter, and so on. When the vegetation is composed of an annual plant, a single measurement of the standing crop at the time of maximum growth is also a crude measure of productivity. Thus, for many there was confusion between a single standing crop measurement and productivity.

Notwithstanding these problems of conception and definition, productivity research boomed. Terrestrial plant ecologists especially found that measuring production logically followed from their earlier work. These ecologists began amassing the great amount of data required to understand patterns of production over the earth's surface.[26] Ecologists were especially active with these studies in England (Ovington, 1959a, 1959b; Westlake, 1963), Canada

(Bray, 1963; Bray, Lawrence, and Pearson, 1959), Germany (Lieth, 1962; Müller, 1951), Denmark (Möller, 1945; Möller, Müller, and Nielsen, 1954), and Japan (Monsi, 1960; Saeki, 1959; Kuroiwa, 1960a, 1960b; Iwaki, 1958). Plant ecophysiologists also began to examine the factors providing optimum conditions for production and the capacity of leaves to achieve maximum productivity. Bonner's article (1962) in *Science* was especially useful in showing why plants converted solar energy to photosynthate at the surprisingly low efficiency of 1 to 2 percent. The limiting factor was shown to be the low light saturation level of the chloroplasts, which in C3 plants saturates well below full light intensity.

In aquatic habitats, the measurements of oxygen and carbon dioxide uptake or release by the planktonic plants were determined using two sets of bottles. In one set, light was allowed to enter, and in the other light was blacked out so that only the respiration of the organisms was measured. Then the sets were compared. These light-dark bottles are particularly useful in lakes and oceans, and many data on productivity were collected from these habitats. In flowing water, however, the problems of constant mixing and flow and the low populations of plankton make the light-dark bottles less useful. As mentioned, Howard Odum (1956a) capitalized on the peculiar characteristics of flowing systems and developed a technique for measuring the rates of gross primary production, respiration, and gaseous diffusion into and out of the flowing water. Thereafter, he applied this technique to springs, streams, coral reefs (Odum and Odum, 1955), and marine turtle grass beds. His comparisons of data showed that these flowing stream systems were among the most productive anywhere. He credited Ruttner (1953) with showing that a current flowing across a community accelerates the metabolic processes that are limited by a slow rate of diffusion. Although Odum did not especially emphasize the point, it is worth noting that his measurements of flowing water metabolism were measurements of whole systems. Odum was measuring the community as a system, not adding up the metabolism of the components as Lindeman and many others had done. At that time, neither Odum nor anyone else made both kinds of measurements to see if there were a discrepancy. In Silver Springs, Odum devoted his effort to describing the pyramid of organismic biomass, but he did not use these data to infer metabolic requirements nor the productivity of the species. Odum and Odum (1955) used the same approach on Eniwetok atoll. Teal (1957) summed the metabolism of populations for his measurement of the system, but he did not have measurements for the metabolism of the whole system for comparison. Thus, these ecologists were comparing apples and oranges.

Interactions between Components

Lindeman's trophic-dynamic aspect of ecology implemented Elton's food pyramid and made the ecosystem coincident with this description. The food pyramid model had two immediate values. First, it provided links between the organisms in an ecosystem. Second, it furnished a mechanism to compare trophic levels through efficiency ratios, which could be represented in mathematical terms. These arithmetic ratios attracted Hutchinson as well as many other ecologists. Lindeman had shown that the respiration (or metabolic) costs of energy increased from the autotroph to the secondary biophage level of the trophic pyramid. Further, the ratio of transfer from one level to another (Lindeman called this "progressive efficiency") increased at higher levels of the food cycle. For Cedar Bog Lake the efficiencies were 10 percent for autotrophs, 13.3 percent for primary biophages, and 22.3 percent for secondary biophages.

A variety of ratios were calculated for the data from food chain transfers, and, in general, the patterns found followed those shown by Lindeman (Patten, 1959). Slobodkin (1962), in a review of energy in animals, concluded that the limited data available at that time indicated that the ecological efficiency (output from one level over the ingestion or input to the next level) was between 5 and 20 percent. As we saw earlier, Howard Odum also proposed an efficiency for this transfer ratio of less than 50 percent.

Obviously, in a system where energy is progressively expended for metabolism and is transferred into a heat sink as material is transferred along a trophic chain, at some point there will be no further energy available to maintain the chain. Therefore, ecologists speculated that the most efficient system likely would be a food chain or pyramid with four or five links. Food chains of this length were widely recognized in nature, and it appeared that the energy theory fit the observations. In 1957, however, when studying a food chain (Golley, 1960), I found that as energy was transferred from grass through mice to weasels the ecological efficiencies were 1.2, 0.03, and 0.9 percent. These ratios are unlike those predicted by the Lindeman model and were explained by the fact that these populations were not in a steady state but rather reflected the dynamic properties of an ecosystem. Meadow mice (*Microtus pennsylvanicus*) eat relatively little of the grass available in their habitat, at least at the density studied. The mouse habitat also served as a winter refuge, and mice migrated into the area during the period between January and March. The least weasel (*Mustela rixosa*), a mouse predator, could not consume all of the immigrants. In Silver Springs, where Odum demonstrated that a steady state occurred because of the constant flow rate and water chemistry of the spring, the ratio of energy intake by the trophic levels was 5 percent for autotrophs, 11 percent for primary biophages, 16 percent for secondary biophages, and 6 percent for tertiary biophages. Again, these values do not fit the expected pattern, yet they fall within Slobodkin's prediction of values of between 5 and 20 percent.

The Behavior of the Whole System

I have discussed how a holistic perspective motivated ecologists throughout the early development of ecology and contributed to the development of the ecosystem concept. The Lindeman model was especially important in maintaining a holistic point of view because it provided a way to analyze an ecosystem and to couple the familiar study of populations and communities with the new concepts. Thus, the Lindeman model was integrative, serving a function that Tansley probably had in mind for the ecosystem concept but could not carry out with the experience available to him. Actually, few ecologists appeared to grasp the possibilities of the Lindeman ecosystem concept to study the whole system directly. Part of the problem was there were few methods or models available to guide those developments. Systems developments at that time were almost entirely conceptual. Mathematical applications were beginning to be developed in other sciences but were applied in ecology only in relatively simple population or predator-prey situations (Hollings, 1965; Watt, 1962). Mathematical applications to ecosystems were few, partly because of the limitations of available computing power. The conceptual models provided a way to organize data into a form that represented the underlying systems logic. For example, Howard Odum's (1957) first diagram of an ecosystem (see fig. 4.4) was used to represent Silver Springs but quickly became representative of the energy flow through ecosystems in general (E. P. Odum, 1968). Howard Odum extended this conceptual model to an electrical analog (Odum, 1960), and he also used an analog computer in his teaching and research. The advantage of using an analog computer was that relations between components and flows were represented directly by the wiring circuitry, and therefore one could solve equations representing those relations without doing the mathematics. Howard Odum developed an analog of Ohm's law to represent ecosystem flows. Ohm's law states that the flow of electrical current (A) is proportional to the driving voltage (V) and the resistance (R), which is a property of the circuit:

$$A = 1/R \; V \tag{1}$$

or,

$$A = C \, V \tag{2}$$

where C is the conductivity.

In the more general terms of steady state thermodynamics, Ohm's law is a special case where the flux *(J)* is proportional to the driving thermodynamic flux *(X)* with conductivity *(C)*. That is:

$$J = CX \tag{3}$$

Odum then drew an analogy between Ohm's law and the general equation 3, relating to flux in an ecosystem, and constructed an electrical analog of the structure (see fig. 4.5). Resistances represented the producing and consuming populations, batteries represented solar energy and energy imports, and the flows were currents in the electric wiring. By using variable resistances and switches, the conditions could be varied experimentally. Milliammeters were used to show the flow rates.

The study of complex problems, however, does not move forward entirely through the activities of creative individuals such as Howard Odum. A group of focused minds is also needed. In the mid-1960s, Stanley Auerbach, at the Environmental Program of Oak Ridge National Laboratory, put together a three-man team to develop systems ecology directed toward ecosystem studies. On his staff at Oak Ridge, Auerbach already had Jerry Olson, a student of succession with a theoretical bent. Olson had developed a model made up of differential equations to describe radioactive transfer in a forest (Olson, 1965). He brought George Van Dyne, a former animal scientist with mathematical skills and interests, and Bernard Patten, an aquatic ecologist who demonstrated a systems orientation, to Oak Ridge to form a systems ecology research group with Olson.

Olson, Van Dyne, and Patten were all strong personalities and distinct individuals, but they shared an interest in applying system modeling approaches to ecosystems. They formed one part of a multinodal grouping that created systems ecology as a technical discipline. Besides Howard Odum, another node was formed by Kenneth Watt, in California, and another by C. S. Hollings in Canada. The Oak Ridge group had the advantage of being in a U.S. national laboratory with the technical support of the type needed to organize a systems science and in a leading research organization focused on ecosystem studies, including both terrestrial forests and aquatic systems. The Oak Ridge group also taught a course at the University of Tennessee to train systems ecologists (Van Dyne, 1966).

The personal reflections of one member of the Oak Ridge team, Van Dyne, were written down and published in 1966 and give us an unusual insight into his thinking at this formative time. Van Dyne takes an exceptionally syncretic view of ecology and systems ecology, seeing systems ecology—as synonymous with ecosystem ecology—as lying at the end of a continuum with physics at the other end. It applies mathematical tools to the description of interactive flows between ecological units and subunits. Systems ecology in 1966 mainly used deterministic first-order equations, but Van Dyne saw advances in computers as opening up dynamic analysis in new ways. He also saw systems ecology as requiring a new kind of scientist, one able use mathematics, having a synthetic mind, and willing to deal with applied problem solving. Finally, Van Dyne emphasized the need for teams of ecologists to study ecosystems and provide data in the form required for systems models of ecosystems. Systems ecology was not merely the outgrowth of systems or operations research and cybernetics. It was necessary because ecosystems were usually too large to be experimental objects. With an ecosystem model however, ecologists could pose questions of ecosystems and test their predictions. Systems ecology established a way of validating ecosystem theory.

Application of Ecosystem Concepts to Ecology

During this period of active growth and transformation, the ecosystem concept influenced other areas of ecological science. Important interactions occurred on topics such as selection, adaptation, stability, and succession, each representing the response of the biotic components of ecosystems over time.

Studies of selection and adaptation were of major interest to ecologists and were becoming a research area rivaling ecosystem studies. Organisms require energy for maintenance and production and must compete with one another for these essentials within food webs. Competition leads to adaptive relationships between populations and toward the evolution of mutually surviving forms within the community—otherwise a particular population would become extinct. Of those ecologists active in the development of the ecosystem concept and ecological energetics, only Slobodkin consistently brought the evolutionary arguments that Tansley had anticipated into his discussion. For example, in his examination of the caloric content of organisms, Slobodkin framed the three alternatives for the distributions of caloric values in selectionist terms. He suggested that a skewed distribution of caloric values indicated that energy was limiting, and further that it was of selective advantage to an organism to use the energy available to increase the number of offspring rather than to increase body energy through the storage of fat. In his analysis of evolution, Slobodkin (1964, 355) stressed flexibility: "From a consideration of the relation between physiological, ecological, and genetic mechanisms by which populations adjust to their environments it seems likely that only flexibility or homeostatic ability is always maintained at a relatively high level by the evolutionary process."

Slobodkin's almost unique position was due partly to his studying populations of organisms enclosed in laboratory vessels where adaptation to the environment and selection could be observed or inferred. Also, he was working within a research community that was becoming increasingly interested in evolutionary questions. Other active ecosystem research groups were less interested in these questions of the relations between populations. Rather, they focused on describing ecosystem structures and functions, searching for physiological mechanisms to explain the patterns they observed.

Stability pertains to the state of the ecosystem over a period of time. Most ecologists accepted that it was necessary to examine a population or community when it was in a steady state. It was only then that the relations between the energy parameters were consistent over a period and measurable as rate processes. Otherwise, the ecologist was chasing variables. But what is a steady state? Howard Odum (1957, 108) argued that Silver Springs was a steady state system. He defined homeostasis as the spring having a constant structure of vegetation and water, a consistent water discharge rate and chemical composition, and a balance in the organic matter cycle over a year. Odum stated that "where all of the components of a community turn over several times a year there would be ample opportunity for changes to occur if there were no self regulative mechanisms. In Silver Springs there is apparently a fairly high degree of stability." He suggested that the stability of the system can be measured by the number of times it turns over without change. Silver Springs turned over eight times in a year.

Of course, short-term irregularities occur in all systems. In Silver Springs, a summer pulse of primary production due to greater solar energy available to the plants led to an increase in the reproduction of consumers but not to a change in their standing crops. Odum interpreted this observation as an example of system regulation, whereas Slobodkin used this same type of observation as evidence of the role of evolution in structuring the caloric values of organisms. Thomas Brock (1967) suggested that while short-term instabilities can be ignored in evaluating the steady state, we can also ignore long-term evolutionary changes that are superimposed on the steady state. He observed that a forest may be changing slowly over hundreds of years, but when viewed on an annual basis, it is in a steady state.

Stability, however, involves more than a balance of energy flow and production. The stability principle stated: "Any naturally enclosed system with energy flowing through it whether the earth itself or a smaller unit, such as a lake, tends to change until a stable adjustment, with self-regulating mechanisms, is developed. Self-regulating mechanisms are mechanisms which bring about a return to constancy if a system is caused to change from the stable state by a momentary outside influence. A governor on a steam engine is an example of a self-regulating mechanical mechanism" (Odum, 1959, 45–46). Self-regulation provides the mechanism to achieve the equilibrium state postulated by Tansley. What was the mechanism—the governor—in real ecosystems?

The answer to this question came in two forms, which as so often happens, were interconnected. First, comparable to systems engineering, the control was produced by feedback. Feedback is a process that has a positive or negative effect on another process. For example, herbivores feeding on leaves can have a depressing effect on primary production because they remove the structures that are accomplishing photosynthesis. In this case, the feedback is negative and production will decline with higher feeding rates. Feedback may also be positive and increase the rate process. For example, a herbivore's manure provides nutrients required by the plant for its production. Since it takes time for the feedback effect to take place, there is a time lag for feedback control to operate. Stability is maintained by feedback control, and the process being controlled oscillates at a frequency that is dependent on the time lag.

The second part of the answer came primarily from the work of Ramon Margalef and Robert MacArthur. Margalef (1963) observed that as ecosystems became more mature, there was an increase in structural complexity and a decrease in the energy flow per unit of biomass. In mature systems, he said, "the future situation is more dependent on the present than it is on inputs coming from outside. . . . Maturity is self-preserving. . . . Fluctuations in more mature ecosystems are more dependent on internal conditions of equilibrium, that is, on biotic factors." Margalef's main point was that stability is related to biotic complexity. Less energy is required for stability in a more complex system, and the trend with maturity is toward a decrease in the energy flow per unit of biomass and an increase in organization. This involves a natural selection where the links between components in an ecosystem can be substituted for other links that work with higher efficiency, saving potential energy that can be used to develop the system further.

MacArthur (1955, 534) also contributed to the idea of the selection of energy exchange links being related to stability. He graphed a food web containing four species and showed that the transfer of energy conformed to a Markov chain in probability theory. Reasoning from this food web to the stability of species in communities, MacArthur concluded: "The amount of choice which the energy has in following paths up through the food web is a measure of the stability of the community." He cited Eugene Odum (1953) as the source for this idea. The complexity of the food web creates alternative flows of energy, allowing selection for more efficient species in energy transfer and for a system to remain in balance as a whole. Simple systems with simple food webs would be less able to cope with biotic changes that might disrupt an efficient transfer of energy. Thus, the stability of an ecosystem depended directly on the diversity of species within the system. The species represent different food niches and transfers of energy.

The idea that diversity produces stability was supported initially by observation. For example, Eugene Odum (1960) and I (Golley, 1965) showed that an increase in the number of plant species in abandoned fields undergoing plant succession was correlated with the stability of the production rate. This idea also was developed in an important conceptual monograph by Joseph Connell and Eduardo Orias (1964). The authors stated that stability might be observed in simple communities, such as salt marshes, when the physical environment created conditions for growth of a stable community.

Eventually the concept was examined at a workshop at Brookhaven National Laboratory, New York (Woodwell and Smith, 1969), and was found wanting. Simple systems may be stable, and species-rich communities may be unstable. No universal pattern holds. Nevertheless, the environmental movement of the late 1960s and 1970s used the diversity-stability hypothesis as a central tenet supporting conservation action, and it is still being taught as a common sense relation. Carl Linnaeus's balance of nature concept (Egerton, 1973) remains alive and well in the popular mind. Possibly, ecosystems are never stable but are always in a process of change. That is a question for later consideration.

Ecosystem concepts continued to be applied to studies of ecological succession (Odum, 1969). There is an obvious connection between the concept of a steady state and the concept of climax. Tansley, at the 1926 International Congress of Plant Science (Tansley, 1929, 686), commented on climax in the following terms: "A climax community is a particular aggregation which lasts in its main features and is not replaced by another, for a certain length of time; it is indispensable as a conception, but viewed from another standpoint it is a mere aggregation of plants on some of whose qualities as an aggregation we find it useful to insist."

Although Tansley's expression was a step away from the Clementsian concept, it was not until 1953 that distinguished ecologist Robert Whittaker presented a modern view of ecological climax:

> 1) The climax is a steady state of community productivity, structure and population, with the dynamic balance of its populations determined in relation to its site. 2) The balance among populations shifts with change in environment, so that climax vegetation is a pattern of populations corresponding to the pattern of environmental gradients, and more or

less diverse according to diversity of environments and kinds of populations in the pattern. 3) Since whatever affects populations may affect
climax compositions, this is determined by, or in relation to, all "factors"
of the mature ecosystem—properties of each of the species involved, climate, soil, and other aspects of site, biotic interrelations, floristic and
faunistic availability, chances of dispersal and interaction, etc. There is no
absolute climax for any area, and climax composition has meaning only
relative to position along environmental gradients and to other factors."
(Whittaker, 1953, 61)

Thus, Whittaker presented a modern concept of ecological succession and
climax that could be used to interpret the actual patterns of succession observed
in the field within ecosystem theory.

One of the principal groups studying ecological succession from an ecosystem perspective was that of Eugene Odum. The group's research was done
mainly on the property of the Savannah River plant.[27] This plant site was
located on the sandy upper coastal plain in South Carolina bordering the
Savannah River. The land was purchased from private citizens, and they were
required to vacate the property in 1951 and 1952 when construction of
reactors and support facilities began. Odum and his students were faced with
the almost three hundred square miles of abandoned land that was undergoing
ecological succession all at one time. He applied the ecosystem approach
directly to the problem of understanding this vast area, focusing on the productivity of vegetation and the dynamic behavior of selected groups of consumers.
Odum was limited by money (his initial support was only about $10,000) and
therefore by staff support. For the first few years his research team consisted
only of graduate students. Because the research they carried out was reported in
doctoral theses, the research orientation was on discrete groups of organisms or
situations that could be studied by an individual student. The changes were
described by an imaginative set of long-term, intense, but low-cost observations by the whole team. For example, birds—especially quail and doves
(Golley, 1962)—were counted on transects each year, and lines of steel traps
were operated in spring and fall to record furbearer population trends (Wood
and Odum, 1964).

Odum's own publications for this period included a report on plant succession in the old fields (Odum, 1960). Each year for the first seven years the plant
community changed, gradually becoming more diverse in species and the
number of species sharing dominance. Primary production was highest in the
first years, possibly because of residual fertilizer left in the soil from agriculture.

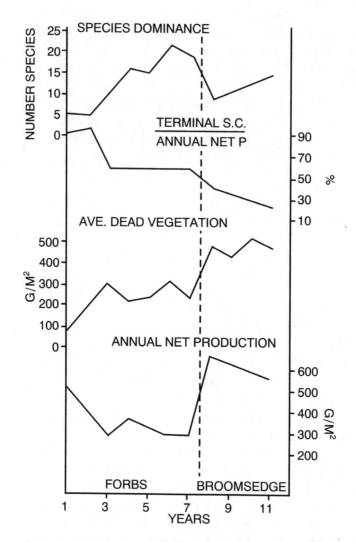

4.8 A comparison of several characteristics of vegetation growing on abandoned fields at the Savannah River plant, Aiken, South Carolina. The diagram shows two stages of old-field succession. The first occurred in years one to seven and was dominated by various herbs. The second began in the eighth year and was dominated by the perennial grass, boomsedge (Golley, 1965). Published with permission of the Ecological Society of America

It then declined and became relatively constant (fig. 4.8). The ratio of the terminal standing crop of vegetation to the plant production also declined over time, from 90 percent in the first two years to about 50 percent in later years.

I continued these studies for four more years, when the community had changed from a herbaceous- to a grass-dominated vegetation (Golley, 1965). I observed the same pattern of change as observed by Odum but with different numerical levels. There was, it appeared, some connection between the number of plant species and the constancy of production. As the vegetation form shifted, a steady state in structural and functional parameters was gradually established and maintained until the next shift in structural form. Presumably, the process would continue until the climax was reached. It was possible to see this type of process operating because the abandoned fields were so large and the invasion of new life forms was slowed. Ordinarily, on small fields of a few hectares size, the process of succession was supposed to be continuous until climax was reached (Oosting, 1942). This observation of a relation between plant species richness, or species diversity, and productivity was part of a research path that occupied ecology for several years.

The George Reserve of the University of Michigan was another active site for ecosystem studies of succession. Here Evans and Hairston and their students studied vegetation (Evans and Dahl, 1955), consumers (Wiegert, 1964), and soil organisms (Hairston and Byers, 1954; Engelmann, 1961). These investigators also took the merological approach, which builds the community from an examination of the parts (Hutchinson, 1964) to a study of the old-field ecosystem.

Theorists noting these observations speculated about succession in an ecosystem context. MacArthur (1955) concluded that the stability of the system was related to the number of possible ways for energy to pass through the ecosystem. Odum and Pinkerton (1955, 342) observed that succession was driven by the amount of production remaining after plant metabolism that was not consumed directly by biophage populations.

> Under these competing conditions the primary producers, the plants, which are best adapted may be the types that can as a group give the greatest power output in the form of growth. According to the argument . . . this should occur when the adjustment of thermodynamic force-ratio of the plants, not of the whole community, R, is 50 per cent. The community of maximum possible size is thus supported. For two climax communities which have similar rates of respiration per unit mass of biological material, the ratios of community standing crops to primary plant productivity should be similar. Thus there is reason to expect productivity and standing crop mass of biological material (biomass) to have a definite relationship under climax conditions. Communities which do not have the maximum biomass would pass through successive generations until they achieved this condition.

Finally, Margalef (1963, 361–62), in an important treatise, brought together many of the theoretical ideas relating to succession in ecosystems. He emphasized the ratio of primary production to total biomass (in his terms, "the keeper of organization") of the ecosystem, which is lower in mature systems with richer structure. In the more mature systems, there is a more complete use of food, a greater proportion of the animals, and energy cascades through more steps. In succession the two most notable changes are an increase in complexity of structure and a decrease in the energy flow per unit of biomass. Margalef drew an analogy to selection: "Links between the elements of an ecosystem can be substituted by other links that work with a higher efficiency, requiring a change in the elements and often an increase in the number of elements and connections. The new situation now has an excess of potential energy. This can be used in developing the ecosystem further, for instance, by adding biomass after driving more matter into the system. A more complex state, with a reduced waste of energy, allows maintenance of the same biomass with a lower supply of energy—or a higher biomass with the same supply of energy—and replaces automatically any previous state."

Thus, the empiricists and theorists both adopted Tansley's conception of a climax community as an aggregation that lasts for a certain length of time, and they interpreted ecological succession within the ecosystem concept. It is important to point out that this was not a transference of Clementsian ecology to ecosystem studies, as alleged so frequently by later ecologists. None of the ecosystem ecologists studying succession cited Clements as a source for their ideas. Rather, the point of similarity was that they, like Clements, accepted the reality of an object consisting of many different species. Aggregation, community, and ecosystem are all concepts based on a recognition of an object as made up of species populations. The questions addressed by ecosystem ecologists were, How do these ecosystems change over time? Do they maintain a steady state? Through these questions, studies of ecological succession were given a new, modern context and contributed to a larger, more modern theory.

FINAL WORDS

During this period of about fifteen years, the ecosystem concept became established as a scientific paradigm in ecology. This paradigm described an ecosystem as an ecological machine constructed of trophic levels that were coupled through flows of energy.[28] Ecosystems were in steady state when the input and output of energy was balanced and no accumulation of biomass through productivity was observed. If there was an accumulation of energy and materials, as in successional ecosystems, the system would expand in biomass and species diversity until a steady state was reestablished.

This paradigm was articulated most clearly by Eugene Odum in his textbook, *Fundamentals of Ecology*. It was explored by a growing number of scientific workers, many educated by reading *Fundamentals*. The paradigm dominated general ecology, as distinct from plant ecology and animal ecology, which continued development along other lines. One way to demonstrate this relation is through a count of the titles listed by *Biological Abstracts*, which began indexing the term *ecosystem* in 1957. The number of articles with the word *ecosystem* in the title increased from a single one in 1957 to 162 in 1975 (see fig. 4.4). This number does not include all of the articles on ecosystems, and especially misses those concerned with process studies in plant or animal populations. The growth in ecosystem titles can be compared with articles indexed under the term *ecology*. Ecological articles increased from 37 to 1,759 in the same period. These data show that ecosystem studies increased with ecological studies generally, sharing in the growth of scientific activity following the second world war. Many ecosystem researchers also carried out studies in plant and animal ecology, and there was substantial interaction between all types of ecologists, who at this time represented a rather small group of specialists.[29]

Research on ecosystems occurred mainly in America, largely because of funding by the U.S. Atomic Energy Commission. The officials in the AEC were able to visualize several direct applications of ecosystem theory to applied problems. The AEC was concerned with the transport and fate of radioactive materials produced by weapons testing, the response of biological systems to ionizing radiation, and with establishing baseline conditions at their nuclear energy facilities. The emerging ecosystem studies were directly relevant to those needs, and the AEC provided relatively large sums to ecologists to organize research groups and study problems. These groups gradually became permanent organizations carrying out a variety of ecological studies at the AEC National Laboratories and production sites. As a consequence, radiation ecology became an applied area of ecosystem studies. In other countries the expansion of ecosystem research was much less active. Only in the study of organic production were European and Japanese contributions equal to or greater than those of Americans. The relation between ecosystem research and the military activity of the United States was never obvious, and ecologists seemed oblivious to the connection.

The theoretical aspect of the paradigm also developed during this period. Tansley's concept of a system made up of the "organism complex" and the "environment of the biome" had been applied to a real object, a lake, by

Lindeman. The problem of whether an ecosystem was a concept that could be applied to ecological systems of all sizes and complexities as Evans (1956) had asserted or was applicable only to a specific type of ecological system consisting of a bounded biotic community and its environment, as Lindeman had done, was ignored. Ecosystem theory developed along three broad avenues. First, concepts of structure and function of whole ecosystems were developed largely by the Odum brothers and Margalef. Concepts of energy flow in ecosystems were developed by Howard Odum, Lawrence Slobodkin, and his students, Richard Wiegert, the students of Eugene Odum, and myself. Finally, plant ecologists and limnologists continued to expand the theory of organic production, while animal ecologists developed concepts of energy flow through animal populations. By the mid-1960s, many researchers were contributing to the description of ecosystems and their functions.

In this post-Lindeman, post–second world war period, ecosystem research remained mainly descriptive. Howard Odum's study of Silver Springs was the paradigmatic example of these descriptions of entire ecosystems, replacing Lindeman's Cedar Bog Lake. Theory developed alongside description, as ecologists proposed principles of ecosystem structure and function that were analogous to the principles governing physical and information systems. The description of whole systems and the theory provided an exciting but ill-defined and poorly integrated body of science. An operational principle or method that would organize the data and concepts was missing, although advances in systems analysis suggested a potential direction for ecosystem studies. An apologia for ecosystem studies was expressed most vigorously by Eugene Odum, who defended the concept against attacks by reductionists. Odum asserted that the ecosystem was a whole greater than the sum of its parts and had emergent properties that could not be explained solely from a knowledge of the parts.

Thus, by the mid-1960s, there was a well-established, active scientific community engaged in applying a paradigm that closely fit the social-cultural environment of postwar America. Fundamentally, it was a mechanical and economic perspective of nature, with energy as the currency of exchange between components. Because there was no method available to keep track of the exchanges between all species in an ecosystem, it was necessary to aggregate species into functional groups, the trophic levels. This process simplified the task of organizing data and opened the door to the use of systems analysis in ecosystem studies. But the choice of aggregating organisms into trophic levels was a double-edged sword. Although the advantages were many, the disadvantage was that most of the biological reality encompassed in the species was lost. In the ecosystem model, species acted abstractly, like robots. This decision cut ecosystem studies off from biology and natural history and linked them more closely to engineering, physics, and mathematics. Even though this was a desirable and prestigious link, the costs were also large.

The choice of the trophic level model meant that the field could move more quickly, it did not become bogged down in biological detail, it was not retarded by an ignorance of feeding biology, and it could profit from advances in systems engineering, information, and computer science. It fit remarkably well with the spirit of the time. The word *ecosystem* expressed this spirit. I think we can see a juxtaposition between scientific advance and popular interest and need that accounted for the success of the concept.

The most important point about this step was that for the first time ecologists identified a specifically ecological object for study. Although the ecosystem contained biological species and physical and chemical processes in the environment, ecologists moved from being biologists, chemists, and physicists to being members of a new scientific discipline. Eugene Odum proclaimed the "new ecology," yet remained ambiguously an ecologist and a zoologist. Few of his colleagues went even that far, continuing to interpret ecology as a subdiscipline of biology in its traditional form. Yet the step had occurred, and the public had a concept that could be used to bring together information from diverse disciplines and apply it to environmental problem solving.

The weakness or strength of a scientific concept is usually not crucial, because it is soon revised as scientists apply it in their work. In this case, however, the concepts were not presented in the conventional scientific form. Rather, they were derived from authority figures who frequently were the professors of the key investigators. The ideas were often presented in authoritative language and, most important, as principles in the textbooks used to train the next generation. There was little dissent. The consequence was rapid movement in a single direction. We can observe this as growth in the richness of the paradigm. This richness increased as it was applied in radiation ecology and environmental management.

By the end of the era, however, a number of critics began to point out problems in ecosystem studies. The ecosystem research community, intent on describing the structure and function of ecosystems or confirming concepts, was unable or unwilling to respond to the criticism. Rather, ecosystem ecologists tended to restate the ideas underlying the field and further define their concepts. Eugene Odum was most articulate in defense of the viewpoint and became, as a consequence, the acknowledged leader of ecosystem studies internationally.

In retrospect, I think the fact that the ecosystem concepts were not presented as hypotheses to be tested or questions to be asked was the most serious weakness of the science in this period. The growth, excitement, and advances were logical developments of the founding ideas coupled by analogies to other forms of systems. It fit the time. Ecologists were not questioning the cultural paradigms, they were working within them. It was not until later, when authority was generally questioned in American culture, that ecosystem theory came under an increasing amount of criticism that could no longer be ignored.

CHAPTER 5

The International
Biological Program

By the mid-1960s ecosystem studies had reached a level of worldwide activity. Ecosystem studies were a popular, even dominant, form of ecological research. While teams of researchers at national laboratories and other centers studied whole forests or lakes, individual researchers examined processes within systems, such as the rates of primary production, transfers between trophic levels and populations, and the rate of organic decomposition. An unorganized body of theory was available to stimulate research. Eugene and Howard Odum, Margalef, Slobodkin, and others viewed ecosystems from a variety of perspectives, frequently reasoning analogically from physical, chemical, or biological systems to ecosystems. The condition of ecosystem studies at this time might be characterized by Claude Levi-Strauss's term *bricolage,* which refers to the construction of an object or a theory from a variety of unrelated, found materials. The *bricoleur* arranges these and creates something new and unexpected from the disparate materials. Ecosystem theory was constructed from thermodynamics, from physical equilibrium theory, from information theory, from evolutionary theory, from field natural history, and so on. In 1965 it did not yet form a coherent, organized body of knowledge.

It was suggested by several systems ecologists that the way through this stage of science development was to employ systems theory to design models to organize the information about ecosystems. Olson, Patten, and Van Dyne at Oak Ridge were most enthusiastic about this approach, and they predicted a substantial advance in both the understanding and utility of ecosystem theory using systems ecology. Actually, the discipline faced two alternatives, although the choice was not clearly recognized by the community at the time. On one side was the system ecology approach, which constructed ecosystem models from the information about its components and linkages. These models represented the ecosystem type in a general way and would prove useful in predicting ecosystem performance under changing environmental conditions. On the other side was the approach pioneered by Howard Odum at Silver Springs. The ecosystem was considered as an object of research. Its input and output properties were determined and then the components were examined to explain the conversion of inputs into outputs by the system. In the next ten years, these two alternatives were tested in a scientific experiment carried out mainly in the United States. The opportunity for the test was an international scientific program called the International Biological Program. In this chapter I present the IBP from the perspective of its impact on ecosystem science.

Only one part of IBP, biome studies, was focused on ecosystem studies. *Biome* was a word coined by Frederic Clements to refer to broad, regionwide associations of plants and animals, such as tundra, boreal forests, and so forth.[1] The size and complexity of the IBP biome program as it existed preclude a detailed examination of each project. Rather, my strategy will be to focus on one program—the grassland biome project—and then to contrast selected features of the other studies within that project. In this way we will be able to identify some of the key factors that influenced ecosystem science and that led, or did not lead, to a new stage of development.

THE DEVELOPMENT OF THE INTERNATIONAL BIOLOGICAL PROGRAM

The International Biological Program became a major vehicle for ecosystem studies, but when it was first conceived by Sir Rudolph Peters, an English biochemist and president of the International Council of Scientific Unions (ICSU), as a biological imitation of the International Geophysical Year (IGY), he visualized a project in nucleic acids (Peters, 1975). The idea was first put forth in March 1959 after a meeting of the ICSU executive committee in Cambridge, England, when Peters, Lloyd Berkner, an American physicist and past president of ICSU, and Giuseppe Montalenti, president of the International Union of Biological Sciences (IUBS) and a geneticist, were returning to London on the same train. The idea spread rapidly, and in 1960, at the ICSU's next executive committee meeting, Montalenti presented a formal scheme for the IBP (Waddington, 1975). The overall theme chosen was "The Biological Basis of Human Welfare." A preparatory committee was formed, which met in March 1961 and drew up a proposal with three project areas: human heredity; plant genetics and breeding; and studies of natural communities that were liable to undergo modification or destruction. At the IUBS's general assembly meeting later that year, the committee's proposals were considered and a resolution was passed affirming that the program's aim would be "toward the betterment of mankind." The three specific areas for action would be conservation, human genetics, and improvements in the use of natural resources.

The planning committee met again in May 1962 in Morges, Switzerland, at the headquarters of the International Union for the Conservation of Nature (IUCN), where the final organization of the IBP was developed. Seven sections were organized. Three focused on terrestrial productivity; these included general productivity, metabolic processes, and the conservation of threatened communities. A fourth section was concerned with productivity in fresh water and a fifth with productivity in marine environments. A sixth section dealt with human adaptability both physiologically and genetically, and the seventh, titled "Use and Management," focused on aspects of applied biology outside of the interests of the World Health Organization (WHO) and the Food and Agriculture Organization (FAO) of the United Nations. The program was launched at the first general assembly of the IBP in Paris in 1964. Here it was resolved that the program should contribute to "the optimum exploitation, on a global basis, of the biological resources on which mankind is vitally dependent for its food and for many other products." The program was to have a finite life, terminating in 1972, but was later extended to 1974.

A book on the origin and development of the IBP (Worthington, 1975) follows the twists and turns that were required before the final compromise program was designed. Unlike the geophysicists, the biologists did not have a material global object on which to focus a global program. Biologists were required to create a global purpose from abstract needs and concepts. Peter's nucleic acid idea was quickly dropped, because he concluded that the field was "fully stretched." Montalenti's interest in human populations, in contrast, was converted into a program on human adaptability. Academicians Andrey Lvovich Kursanov, a plant physiologist, and Vladimir Aleksandrovich Engelhardt, a biochemist, the Russian members of the ICSU executive committee and the ICSU planning committee, pressed to have the program deal with the rational use of plant and animal resources with the aim of raising the standard of human life. This theme was reflected in the program's title and by its emphasis on productivity. Conrad H. Waddington thought that the study of the way solar energy is processed by the biological world to form complex molecules, which are used as human food should be a theme of the IBP. A conservation theme had been proposed early in the planning.

Thus, the themes of conservation, human adaptability, and use and management were grouped with an overall focus on biological productivity as a basis for human well-being. In retrospect, this focus made good sense. Productivity was a scientific concept that had been well developed by the early 1960s. It would attract scientists in agriculture, forestry, and fisheries, as well as those in basic ecology and fundamental studies in plant and animal biochemistry and metabolism. Also, it was broad enough to cover the different interests of the various national members.

The IBP was a complex program with projects in many areas of biology. Ultimately, however, it developed into a largely ecological program. The production part was developed in two phases. The first phase, from 1964 to 1967, was characterized by a series of planning meetings, the development of methodology handbooks, and the beginning of some limited research projects. The second phase, from 1967 to 1974, was marked by a growing dominance of the biome programs, which gradually became so active that the name IBP became in some people's minds synonymous with biome studies. The scientific director of the IBP, E. Barton Worthington (1975, 64), writing in the synthesis volume, states the matter well:

> It is this ecosystem approach which distinguishes much of the IBP research from what had dominated ecology before. Essentially it consists of the careful selection of a number of variables—biological, chemical and physical—about which data are collected, quantitatively as well as qualitatively. Thereby the ecosystem can be analyzed in order to ascertain which factors and processes are important in causing the dynamics of the whole. In this, the application of system analysis to biological systems has been one of the major innovations developed during IBP. Some would go further and say that it has been one of IBP's major achievements.

International programs do not evolve in a vacuum. The organizers must have good reasons for devoting the immense amount of time and effort required to plan a program and to convince fellow scientists and funding sources that the work is needed and will yield productive results. The original impetus for the IBP was, apparently, the desire to obtain some of the same benefits that had accrued to the earth sciences from the International Geophysical Year. These included the creation of international networks of collaborators, the collection of data on large-scale systems, and obtaining the attention of decision makers.

In addition, large communities of scientists were already at work, or were capable of beginning work immediately, and were eager to organize internationally. The freshwater biologists, for example, by 1962 had presented a report

5.1 Participants at the IBP Terrestrial Production project organizational meeting in Jablonna, Poland, 1966. From left to right, F. B. Golley, R. G. Wiegert, D. A. Crossley, Jr., Kazimierz Petrusewicz, Alicja Breymeyer, and Francis Evans. The man in the rear is unidentified. Photo property of the author

on the opportunities for limnology in the IBP,[2] prepared by Wilhelm Rodhe, a Swedish limnologist. This report was considered at the fifteenth Congress of *Societas Internationalis Limnologiae Theoreticae et Applicatae* (SIL). The marine biologists were also thinking globally and had several international programs under way.[3] The IBP, in their case, focused on topics not adequately covered in these other programs. For these fields and for the conservationists, IBP provided an opportunity to build better linkages between national communities, collect data globally, and to look for patterns on a global or regional scale.

The terrestrial group began with a meeting in September 1963, organized by Paul Duvigneaud, a plant ecologist and biogeochemist from Belgium. Attending were mainly plant ecologists and vegetation scientists interested in global patterns. Yet the official orientation of the terrestrial scientists did not continue in this botanical direction. In 1966, a meeting on the secondary productivity of terrestrial ecosystems was convened in Jablonna, Poland, under the leadership of François Bourliere, a French mammalian ecologist and professor of medicine in Paris. This meeting included many scientists working on ecosystem processes (fig. 5.1), and after long discussions,[4] the participants

Species Groups	Habitats Tundra	Boreal Forest	Temperate Forest	Grassland	Tropical Forest
Vegetation					
Small mammals					
Large grazing mammals					
Granvorous birds					
Social insects					
Soil organisms					

5.2 A matrix developed to organize research in terrestrial productivity of the International Biological Program, approved at the Jablonna, Poland, organizational meeting in August 1966. Groups of organisms were to be compared within and across habitats.

agreed to a program designed as a matrix (fig. 5.2). At the top of this matrix were placed the major ecological formations of the world, indicating the principal habitats of animals. Emphasis was on temperate regions, although the tundra, taiga, and tropical forests were included. The rows of the matrix consisted of groups of organisms. Over Duvigneaud's objections, vegetation was treated as a group, as were insect herbivores, social insects, small mammals, large grazing herbivores, granivorous birds, and so forth. In this way, terrestrial productivity included within the IBP some of the subdivisions of biology focusing on taxa, along with ecological subgroups interested in the properties of ecosystems, such as the flow of energy, productivity, and the decomposition of organic matter. This plan went beyond a focus on secondary productivity,[5] since it included vegetation as part of the matrix and treated it as the primary producer component of the ecosystems.[6]

This scheme represents the biologically oriented ecologists' definition of a global program. The emphasis was on groups of organisms studied across habitats and regions. It took a biogeographic approach to the problem. Processes were not emphasized, except as they were represented by organism groups.

THE U.S. PLAN

The Jablonna matrix lasted as the accepted plan for all IBP terrestrial studies for about one month. Ecologists in the United States met about a month after the

Jablonna meeting at Williamstown, Massachusetts, and developed an IBP of their own, which over years gradually influenced and changed the international focus.

The United States was part of the IBP almost from the beginning. George Ledyard Stebbins, secretary general of the IUBS and a Davis, California, plant geneticist, offered his ideas about potential objectives at the first meetings to organize the IBP. He had queried American biologists about their interest in forming an IBP, although with not very useful results,[7] according to Waddington (1975). When the special committee for the IBP was formed by ICSU, Stanley Cain, a Michigan botany professor and at the time (1965–68) assistant secretary for Fish, Wildlife, Parks and Marine Resources of the U.S. Department of the Interior, was chosen as a member. Arthur Hasler of the University of Wisconsin, George K. Davis, a Florida nutritionist and biochemist, and Bostwick H. Ketchum from the Woods Hole Marine Laboratory in Massachusetts were also involved in planning from about 1964. The U.S. National Academy of Science (NAS) provided a special IBP planning grant of $50,000 a year for four years, from 1965 through 1968. Even so, Waddington felt that the American community of biologists was the most difficult group to have become involved in the program. Waddington commented: "The idea of studying the energy balance of ecosystems had therefore got to thread its way between adherents of the 'central dogma,' who couldn't care less but were apprehensive it might take away some of their public funds, and an opposite party whose line was that 'we *are* field biology, and productivity is not an American problem'" (Waddington, 1975, 9). In 1963, after a meeting in Washington, D.C., arranged by T. C. Byerly, a U.S. Department of Agriculture animal biologist, Waddington was sufficiently disturbed by the hostility of senior American biologists to participation of the United States in the international program that he wrote to a number of his friends, such as C. H. Müller, Tracy Morton Sonnenborn, Sewall Wright, Ernst Mayr, Theodosius Dobzhansky, and James Ebert, all distinguished American biologists, asking for advice and support. The response from Ebert was that a well-designed international program would be supported (Waddington, 1975). This support eventually resulted in a financial grant from NAS and in the organization of IBP planning committees.

U.S. IBP ACTIVITY BEGINS

Although Waddington's concern about U.S. participation in the IBP may have been correct in 1963, by 1964 interest had begun to burgeon. The U.S. delegation to the first General Assembly in Paris consisted of nineteen people, with T. C. Byerly and S. A. Cain as cochairs. The delegates represented, in addition to the official U.S. group, the Conservation Foundation, the Nature Conservancy, the IUCN, the National Science Foundation, the Ecological Society of America, the IUBS, and several ICSU special committees. At the Paris meeting, James B. Cragg, an English zoologist and chair of a committee including Heinz Ellenberg from Germany and forest ecologist J. D. Ovington from England, proposed the formation of a production terrestrial (PT) group. Their proposal included the following statements: "Nevertheless studies of organic production and decomposition at different trophic levels seem to provide a unifying factor and should be a fundamental part of every project in terrestrial ecology. Analysis of dry matter can provide basic data on energy flow, protein production, mineral cycling, environmental pollution, etc. The interrelationships between production and biological diversity, community structure and the living organisms of different communities can be examined as well as environmental factors such as climate and soil."[8]

The U.S. delegation returned home and began the development of the program. A U.S. national committee was formed, chaired by Professor Roger Revelle of the Harvard Center for Population Studies.[9] Revelle was a distinguished scientist, a member of the U.S. National Academy of Science, and a capable organizer. He was assisted by Byerly and Cain as cochairs. Subcommittees or subsections corresponding to the IBP themes were also formed. The terrestrial productivity group was chaired by Eugene Odum, freshwater production (PF) by Arthur Hasler, and marine production (PM) by B. H. Ketchum. In late 1965, Donald Hornig, director of the Office of Science and Technology of the U.S. government, wrote Leland Haworth, director of the U.S. National Science Foundation, asking NSF to coordinate an interagency committee on the IBP. Haworth appointed Harve Carlson, director of the NSF Division of Biological and Medical Sciences, as the chair of this coordinating committee. In 1965, the subcommittees began planning their research programs.

The U.S. IBP organization had the same problem as the international organization in determining how to design a program that would be attractive to all biologists. Although the proposed program initially was broad, including various studies on topics of interest to U.S. biologists, one effort focused on ecosystems directly. The Analysis of Ecosystems program grew from a series of meetings of the subcommittees concerned with production ecology.

The first meeting of the terrestrial productivity group took place in May 1965. The participants included Robert Whittaker, whose most recent work had been at the ecosystem study site at Brookhaven National Laboratory in New York; Larry Bliss, a plant ecologist from the University of Illinois with experience in the tundra; Frank Pitelka, from the University of California at

Berkeley, who was also interested in tundra; and Van Dyne, the modeler and former animal production scientist who was at Oak Ridge National Laboratory. A report was written with contributions from each member. The first draft emphasized research on productivity, training ecologists to carry out the IBP studies, and interaction with scientists in Latin American countries.

As chair, Eugene Odum worked on this draft report and gradually forged a document that was designed to be a national program statement. His report had two special features. First, through contact with the freshwater production committee, Odum's group had developed the idea of focusing research on large areas containing both freshwater and terrestrial habitats. This focus was what ultimately created a very different type of research project.[10] The report proposed as its general objective studying "landscapes as ecosystems," with emphasis on production and trophic structure, energy flow pathways, limiting factors, biogeochemical cycling, and species diversity. It stressed that these studies were not to be confined to natural areas only. The second feature was a proposal to use systems analysis as a mechanism for integrating the results of the study.

Eugene Odum was concerned that the IBP should develop "new thinking at the ecosystem level." In the final program statement, the second guideline was titled "Development of New Methods and Approaches": "Probably the most important role that the PT program can play in the U.S. National IBP effort is that of catalyzing new ideas and techniques which will make it possible to evaluate whole landscapes within the framework of man's dual role as a manipulator of, and a functional component in, ecosystems. The PT program differs from that of the other IBP subcommittees in that it seeks to establish a new science of *landscape ecology* that can provide a 'pure science' basis for landscape planning in the future."[11] This statement was typical of the Odum thought process and expression, and it was prescient.

The terrestrial productivity program statement was to be considered, along with the freshwater production program, at a general meeting of interested ecologists at Williamstown. This meeting was scheduled from 28 to 31 October 1966, and it grew in size until the organizers had to limit the participants based on the availability of hotel rooms. In transmitting the PT program statement, Odum wrote Bliss in July 1966, "Since I have not seen Hasler's final statement, be sure our statement about cooperation with PF in planning, and in hiring a full-time man, jibe with his program statement."[12]

The meeting was a grand success! A new, expanded view of ecosystem research emerged amid a spirit of cooperation and fellowship, helped in no small part by Pitelka, who at the piano led the group in evening song. The final report from this meeting merged the PT and PF committees into a single program focusing on the analysis of ecosystems, with Fred Smith, a professor at the University of Michigan and a theoretical ecologist, as director.

The terrestrial productivity program statement for this new effort said that "the primary purpose of the IBP is an understanding of ecosystems, including man's own." The general objectives of this program were:

1. to study whole systems, such as drainage basins and landscapes, through team effort

2. to study interactions between components

3. to emphasize primary production, trophic structure, energy flow pathways (food chains), limiting factors, interactions of species, bio-geochemical cycling, species diversity, and other attributes that interact to regulate and control the structure and function of communities

4. not to restrict the studies only to natural areas. Ecological succession was to serve as a background in which general objectives could be pursued

5. to consist of collaborative studies in major biomes, in drainage basins, where terrestrial and aquatic studies can be simultaneous

6. to catalyze new techniques, developing theory from small field and laboratory studies

7. to involve systems analysis techniques for examination of existing data on ecosystem processes by sensitivity analysis as an aid in allocating resources for integrated system studies, for rapid organization and analysis of data collected by electronic recording equipment and for analysis and integration of results designed to test and develop theory

8. and finally, to establish centers to store and distribute information collected at the different study sites.[13]

A list of sites in six biomes of North America where work could be carried was also compiled.

Following the Williamstown meeting, Larry Bliss took over the chair of the terrestrial productivity group from Odum. His November 1966 letter to Odum describes Odum's role in developing the PT program.[14] "Your good influence has prevailed in the union of the two subcommittees and in the adoption of the ecosystem theme throughout the program." Yet Bliss was concerned about the communities response. He continued, "I talked with John Wolfe yesterday and he feels we will do well if we can establish one or two good ecosystem studies and that there will be considerable opposition for this program from biologists, to say nothing of other vested interest groups." Wolfe, of course, spoke from the experience of having established the Atomic Energy Commission ecosystem programs.

Nevertheless, the program developed rapidly with a virtual blizzard of meetings in 1967. In February, the U.S. national committee met and approved the plans developed at Williamstown. Also in February a meeting in Chicago considered the criteria to be used for selecting sites for studies. In May, the NSF granted two years of funding for program management, which meant that an office could be established. Also in May a meeting was held on the role of system analysis in the Analysis of Ecosystems program. At this meeting it was decided to focus system analysis first on a grassland site, since the grassland was relatively well understood ecologically and was structurally simpler than forests. In late June, Fred Smith, in a progress report, estimated that the cost of the "Program on Drainage Basins and Landscapes" has been "guessed at $45,000,000."[15] By October, the Pawnee grassland site near Fort Collins, Colorado, had been selected and Van Dyne was appointed director of the project. At this time the first Grassland Working Session was held to design the general scheme of work, which would include both intensive studies at Pawnee site and extensive studies throughout the grassland biome in North America. A smaller group also began discussions on the formation of a tundra study.

Finally, Congress member Emilio Daddario of Connecticut, chair of the subcommittee on Science, Research, and Development of the Committee on Science and Astronautics in the U.S. House of Representatives, held five hearings on the IBP. These hearings included testimony from a number of scientists and were favorable to the program. It appeared that a special funding initiative for the IBP might be provided by the federal government. Philip Johnson, a member of the PT committee and the faculty of the University of Georgia, wrote the Georgia faculty in January 1968 that "the impetus IBP is fostering in ecology cannot simply stop in five years. Ecology will by then have become 'big science' with all the attendant pros and cons."[16] Fred Smith said, "Whatever the causes, a revolution among ecologists is under way, and the IBP is in the middle of it."[17]

The U.S. Analysis of Ecosystems program provided an opportunity for substantial advances in ecosystem studies. It would make a large amount of new funds available to an already well-funded community. It would permit the organization of academic researchers into teams like those at the national laboratories and would focus them on problems of large spatial scale. And finally, it would permit systems ecology to develop its potential organizing and predictive functions.

The Grassland Biome

In October 1967, the Committee for the Analysis of Ecosystems met in Fort Collins, Colorado, and approved a proposed research program for grasslands. The IBP biome programs were under way.

The selection of Van Dyne (fig. 5.3) as director was an important formative factor in the entire Analysis of Ecosystems program. George was thirty-five years old at that time. He was born in southern Colorado into a relatively poor ranching family. As a bright person, he saw the opportunity to better his life with a university education. He began his education at Pueblo College and finished it with a doctorate from the University of California, Davis, in animal nutrition with an emphasis on biometrics and biochemistry. While at the University of California, he taught himself computing techniques by studying and experimenting at nights and on weekends at the computer center. His academic positions included teaching animal husbandry and range management at Colorado State University and Montana State University. He moved to Oak Ridge National Laboratory as an ecologist and was jointly appointed associate professor in the University of Tennessee department of botany.

George Van Dyne brought a familiar American personality type to the IBP: the workaholic. The standards he respected were quantitative, large, and fast. To him, success was largely judged by the amount of money paid—earned or received—from the granting agency; by the number of publications achieved; and by similar measurable criteria. By the time he had reached Oak Ridge, he had authored sixty-seven publications on animal husbandry; according to his bibliography, many of these were reports and experiment station documents. Although many of the publications had coauthors, for almost all he was the senior author. Van Dyne was not inclined to manage an organization by direct personal relations; rather, like the prototypical American business person after whom he patterned himself, he depended on memos, PERT charts, and tables of organization.

Although Van Dyne was selected to be biome director because of his enthusiasm and understanding of modeling, his personal characteristics meant that the IBP grassland program was going to be active in a quantitative way. Van Dyne performed according to expectations. By December 1967 he had organized and sent a 370-page proposal to the National Science Foundation for a grassland program. This work was to extend from April 1968 to March 1969 with a cost for this proposed work of nearly $2 million! The proposal contained the resumes of more than eighty investigators and consultants and over sixty research projects. The work was structured within the Lindeman ecosystem paradigm, with overviews of abiotic factors, and studies of producers, consumers, and decomposers. It contained considerable detail about the ecology of

5.3 George Van Dyne, director of the U.S. Grassland Biome program. Photograph courtesy of Bernard Patten

animal species. Guided by the aims of the national committee, the study included both intensive research at a single site and extensive studies throughout the grassland region. The experimental design and systems procedures were not well developed, but there were organization charts and schedules of work. The proposal ended with a twenty-two-page document on ecological modeling. In this article Van Dyne revealed that he was not interested merely in using modeling to integrate the work, he was interested in creating a new form, or at least, a new application of ecological modeling. This form would be probabilistic and nonlinear.[18]

The initial proposal was not funded. Rather, a revised proposal was requested by NSF. Van Dyne submitted this revised proposal in February 1968 requesting $700,000 for the period April 1968 to June 1969. The reduction in cost was achieved by postponing the start of the comprehensive program, the aquatic studies, and some of the subprojects, delaying some hiring and focusing more attention on the initial modeling. Construction would begin on a building for the project on the Colorado State University campus, and field facilities construction also would get under way.

In June 1968, $400,000 was received from NSF and AEC for the remaining months of the year. This allowed Van Dyne to bring Donald Jameson, a range scientist with a botanical orientation and an interest in system modeling, to Colorado as the manager of the intensive site and to give research contracts to twenty-five scientists. Further, it permitted the program to look beyond the intensive site study toward the comprehensive program. It provided funding for a first meeting of investigators from other grassland sites held at Manhattan, Kansas, in late June.

This was quite an achievement. In nine months Van Dyne and his staff had written two large proposals, involved many investigators across the western United States in the program, implemented the general program plan of the national committee, began to think through the modeling of grasslands, and set up a management system appropriate to a small corporation. It appeared that choosing the grassland as the first biome and Van Dyne as the biome director were the right choices. Clearly, the speed and enthusiasm of the Van Dyne group created a model for other biomes to emulate or reject.

Van Dyne was an enthusiastic spokesperson for ecological modeling. In Washington at NSF, he repeatedly argued that this approach was the way to understand and manage studies of large complex systems. Officials at NSF, especially the assistant director for Biological, Behavioral and Social Sciences, Eloise Clark, were not fully convinced of the ecosystem approach or the utility of models. This was unfamiliar ground, and there were no lack of ecologists who declared that the work outlined could not be done. Van Dyne, in his enthusiasm, overstated his case, and although he obtained support for the program, his exaggerated claims for ecological modeling later came back to haunt him.

In January 1969, Van Dyne, after receiving almost half a million dollars for the period January 1969 to August 1969, submitted another proposal to NSF for the continuation and expansion of the grassland program. This proposal was 675-pages long; in it, he requested $2.2 million for the period September 1969 to December 1970. In the proposal, the full program was described.[19] The broad objectives were to study the "various states of the grassland ecosystems to determine the interrelationships of structure and function, to determine the variability and magnitude of rates of energy flow and nutrient cycling, and to encompass these parameters and variables in an overall systems framework and mathematical model." The research strategy was to employ a systems analysis approach and to isolate and examine the factors and components through the following questions:

1. What are the driving forces making grassland ecosystems operate?

2. What are the major and minor components of grassland ecosystems, and what are the changes in magnitudes of these components over time?

3. What are the important families or groups of processes that cause the interaction, coupling, or linking of the components of the ecosystem one to another?

In the proposal, systems modeling began to receive more attention. Systems analysis is discussed and several important positions taken,[20] including:

A systems approach is the only known method of attaining the objectives of our project.

To our knowledge, no person or group has developed models of entire ecosystems.

Since mathematical modeling of ecological systems is an art in its infancy, we recognize the importance of recording the steps we use in model development and redevelopment. This implies a continued series of reports or publications on stages of modeling.

Most of our participating scientists are not skilled mathematicians. Yet their input into model development is essential. They must recognize the responsibility and accept the challenge.

The functioning of an ecosystem is dependent upon its structure.

These quotations reveal some of the thinking that went into the plans. Much of it was from Van Dyne, and the demand for commitment—because of the significance and value of the exercise—was definitely his. This was the Achilles' heel of the program. Van Dyne was one of the U.S. IBP biome program designers. He tried to follow the plan developed by the national committee in the abstract, and at the same time, create a new kind of program for ecology. He selected the systems model as the key organizing idea, and then with this organizing idea in mind, chose to move toward the program objectives using modern concepts of business management. The resources were not adequate to the task, however, and the people in the project could not be managed as in a private corporation.

The language used in the proposal was a new language for many ecologists. It expressed goals in mechanical systems science terms. How did one translate the word *components* into organisms; *forces* into ecological processes, such as energy flow and mineral cycling; or *coupling* into ecological interactions? Traditional ecologists recruited for the biome programs could interpret the goals in a variety of ways, depending upon the way they decoded this new language.

The National Science Foundation was not willing or able to provide all the funds requested in the several grassland proposals. Within NSF, research is ordinarily funded through a complex program of peer review. The process is linear, with about thirty steps, and involves mail reviews as well as a panel discussion in Washington, D.C. Nonetheless, when a research request exceeds a certain set amount (which was $500,000 during the time of the IBP), additional steps are required. The National Science Board, the policy committee of the foundation made up of the heads of corporations and universities and widely respected scientists, must approve the decision to fund a project. Thus, a proposal the size of the grassland biome proposal had to go through a review process within the ecology program and be approved by the assistant director before being presented to the National Science Board. The board meets at regular intervals, and it requires advanced scheduling to be on the agenda.

One begins to see the complexity of funding this type of work. A proposal might, according to Van Dyne (1972), consist of 100,000 to 200,000 words. It would be reviewed internally two or three times and be rewritten. It would then be reviewed by the U.S. IBP Grassland Biome Scientific Coordinators, an advisory group, then by the Analysis of Ecosystem central staff, and the U.S. IBP Coordination Committee. Only at this point would it be sent to NSF and enter its review system.

The amount of money requested in a proposal was the aggregate from proposals solicited from many individual investigators, adjusted by the Van Dyne management group, and then reviewed and defended up through the committee and peer review system. Although the cost base for the program was determined by many people, the initial request was a compromise between what might be done and what should be done to meet the biome goals, as interpreted by Van Dyne, within the limits of what the participants felt was possible. This was an entirely new kind of venture for ecological science and no one really knew what the National Science Board would do. Therefore it surprised many that the project received about 50 percent of the funds requested in the first case and 80 percent in the second. Still, even this relatively large amount of funding required scaling back the work that was proposed and created management problems. Van Dyne was faced with the prospect of telling people who had been selected to be part of the proposed research group that they would not be funded or that their funding would be much less than required for the work. Further, the funding process was segmented. Van Dyne was required to write a supplement three months after the first proposal was approved in order to obtain all the funds granted for the first period.

Van Dyne (1972, 125), in his no-nonsense, businesslike way, determined that these shortfalls and delays in funding would delay meeting project objectives:

At the outset, we contemplated the completion of a final report in 1,250 working days for a five-year first phase of our program. Yet the calculated duration of the project in the PERT analysis was 1,740 days! Obviously there can be some trade-off between time and dollars. Yet, using our estimates of funding and completion times for activities, it was clear from this analysis this program could not be completed in time. What was feared by many, i. e. that the late start and slow funding in the U.S. IBP, was that IBP would not be able to meet its commitments within the original planned time span through 1972. This preliminary analysis suggested that the final report for the initial phase of our grassland research program could not be completed until between December 1973 and June 1974. Interestingly, and coincidentally, the IBP has recently been lengthened in many countries through June of 1974. We still have a chance!

Not only were the resources inadequate to meet the plan and the method of funding the science Kafka-like, there was also a fundamental problem with staff. The terrestrial ecologists attracted to the biome studies were generally individualists and had never experienced the constraints of a rigid organizational framework. Most were used to the constraints of the classroom or experiment station. Their performance had been judged on the basis of their independent work, and frequently their field work was done alone under arduous conditions. Each competed with others for resources, students, prestige, and recognition. The cooperation that occurred between scientists from the same graduate programs or the same background working on similar problems was frequently expressed in male drinking parties with numerous stories and kidding. The grassland biome proposal and Van Dyne's papers on the project used the male pronoun almost exclusively in referring to scientists. In large part, this was a male world and competition was more the rule.

The biome program assumed that individual scientists could be brought together in a team, focused on a common goal, and that the common goal would be more important than individual goals. That is, the program managers felt that the natural tendency for research to spin off in unpredictable directions could be countermanded through management. Even though the grassland biome began at a time when the popular culture of the United States was supportive of cooperative activity of all kinds, older scientists tended to be less affected by these trends than were students who were caught up into the movements of the day. Although their motivations for becoming involved with the biome program were mixed—a need for funding, interest in a new kind of science and activity, attraction to ecosystem studies, an interest in modeling, or a response to George Van Dyne—people did voluntarily join the project. The problem was to keep them involved, happy, and productive. This proved exceedingly difficult to do.

Van Dyne (1972, 1980) discussed some of the management problems of large-scale ecological work at the same time he was struggling to make the grassland biome program work. These problems included the conflict between the goals of individual research projects and the goals of the model. Ordinarily, a research scientist sets up objectives, collects data, writes an article, and sets new objectives. All the decisions about goals, design, data analysis, and inter-pretation are made by the individual scientist. In the biome program, these steps were open to comment and control by others. An individual's goal might be redefined in a imaginative, exciting way through discussion with a group of peers, but equally it could be altered through coercion. It was possible for a scientist to become only a highly trained technician working for a modeling group. One solution to this problem was to hire students to do most of the actual fieldwork. As a consequence, the quality of the data suffered. Although Van Dyne, from the perspective of modeling activity and the search for trends and patterns, was able to accept some increase in experimental or observational error, the biome scientists were not, and they became concerned about the quality of the research. After an Ecological Society of America symposium in 1972, when these problems were aired publicly, Nelson Hairston made highly critical remarks about the research to the National Academy of Science IBP Coordinating Committee.

Hairston's complaints represented a growing body of opinion in the eco-logical community. Many scientists who were unable or unwilling to join teams began to defend the approach of the individual scientist. Some of the reserva-tions expressed about the IBP when it was first proposed were repeated. The biome programs were believed to be a threat to the individual. It was thought that the large biome research budgets took money from individual ecological research. In a biomelike project, which remained outside of the IBP, led by Gene Likens and Herbert Bormann, the individual nature of the research was emphasized. The high productivity and excellence achieved by their studies at Hubbard Brook were widely used as evidence by others that individual research on ecosystems was more productive than team research.[21]

The debate on how ecological science should be conducted became more virulent during the IBP, and it began to work against ecological science, since decision makers who were not ecologists could not be sure which side was

correct. Young and liberal Americans were attracted to the concept of community and cooperation. Conservative Americans defended individualism. These
inherent differences in political outlook were exaggerated by the debate over
the morality of the Vietnam War: thus, the public enthusiasm for ecology
during the age of environmentalism, which might have been converted to
support new and grandly conceived ecological programs, was frustrated and
diffused. As a consequence, ecological support from the federal government,
including additional funds for the IBP, increased only at the inflation rate
(Golley, 1980). During this same period, support for the other environmental
sciences, where these debates did not occur—such as geology, marine science,
and atmospheric science, increased rapidly at rates well above inflation. Although it is not possible to assign exact causation in this situation, it seems clear
that the debates among distinguished members of the ecological community on
how to best do ecological research had a negative effect on the overall funding
of ecology. In the debates, one side tended to discredit its opponents' position
by pointing out weaknesses in their research designs, and NSF Assistant Director Eloise Clark, a biophysicist, was troubled by the contrary advice she received. Hence, her support for the biome program, which was crucial at the
level of the National Science Board, was not deeply founded.

Van Dyne was an effective competitor. He presented dozens of lectures all
over the United States explaining how the biome program was organized and
what he expected from the research. He and his team wrote voluminously. He
instituted an in-house series of publications, *The Technical Report Series,* which
was meant to inform the research group and eventually included almost three
hundred titles. Articles also began to appear in peer-reviewed journals. A
grassland seminar program moved between the universities in Colorado and
Wyoming, both of which had grassland biome research groups.

In 1969, Norman French was hired as comprehensive site coordinator and
the third member of a three-man management team. French was a mammalo-
gist from the AEC Nevada test site program of the University of California at
Los Angles, with extensive experience studying desert rodents. French developed active programs at about ten other sites throughout the North American
grassland.

Although specialists were brought in to organize and direct detailed studies on their subjects and to organize a modeling team able to take the data and
build models to predict the performance of components, the program lacked
people between the specialists and the modelers able to interpret the data. If this
had been a small project, Van Dyne might have been able to play this role
himself, but he was dealing with hundreds of people. A decision was made to
hire a group of integrators. It brought into the project another group of senior

scientists to add to the team of Van Dyne, Jameson, and French. The 1973 organization chart of the grassland program indicates that these integrators were F. M. Smith, abiotic factors; J. E. Ellis, consumers; D. C. Coleman, decomposers; J. K. Marshall, producers; M. I. Dyer, rate process studies; G. S. Innis, systems analysis; and J. H. Gibson directed services and administration.

Van Dyne worked progressively harder for longer hours, took more trips, and wrote more reports and articles. One staff member recalled that the team received about four memos a day from the director. Yet Van Dyne could not overcome the internal conflicts and contradictions of the program. Don Jameson was the first to leave. In 1971 he received funding for an applied grassland program using a systems analytical perspective from a new NSF program, Research Applied to National Needs, or RANN. Van Dyne, however, did not want to accept a project organized and run separately from the main project. He considered it a deviation from the plan and an expression of unacceptable independence by Jameson. Thus, Jameson left the project, and the task of directing field activities was added to French's other tasks.

Obviously, Van Dyne needed some relief from the pressures of the project, and in 1972 he planned to take sabbatical leave during the next academic year. He planned that he and his wife would visit Europe and spend part of the time in England, where the IBP was headquartered. He went alone. His wife left him, giving him an apparently unexpected blow, which he found extremely hard to accept. Yet the leave seemed to be effective in restoring his equilibrium. He made several trips back to Colorado to check on the program and contribute to the third large proposal, which was being developed under the direction of integrator Dyer, a former Fish and Wildlife Service biologist who was a specialist on the interaction of birds and vegetation. Upon its completion, Van Dyne signed it, and it was sent to NSF, reviewed, and funded.

Upon Van Dyne's return, he reread and reconsidered the proposal and became outraged. He declared it to be counter to the entire pattern of biome development. He immediately went to Washington. At NSF, he demanded that the proposal be withdrawn and that full authority to return the project to its objectives be given to him. His request was denied, and he returned from Washington in a foul mood. His first action when he arrived in Fort Collins was to call the entire staff together and tell them that the project and their employment was reaching its end. He said that he would begin making plans to lay off staff and that they should begin to look for other employment. He labeled this outburst "save six for pall bearers," envisioning a much-reduced program with only six senior staff to work on parts of the large program. One can imagine the consternation this action produced. Here was a group that had successfully written a proposal that produced over two million dollars for each of three years

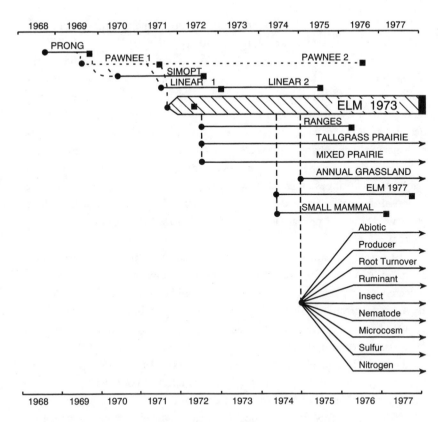

5.4 Evolution of the ELM model of the IBP grassland program. The black boxes represent reports on the model, and the arrows indicate that the work was unpublished in 1978 (Van Dyne, 1978, ix).

being told by their director that the project would end and to start packing their bags.

Today, in rereading these proposals, it is difficult to see what aroused Van Dyne's ire. The June 1973 continuation proposal requested $7.6 million for the period from January 1974 to December 1976, and it continued the earlier development of research in its three hundred pages. There was, however, a discussion about the end of the program: "Our current plans, to obtain the model-experiment feedback and interaction in a more quantitative way, call for detailed modeling, model sensitivity analysis, and synthesis activities in the 1973 and 1974 period to be followed by a final comparative field validation study sequence in 1975 to 1977 [fig. 5.4]. This period would be followed by a final phase of coordinated model development, model experimentation, and synthesis in 1978 and 1979. Although publications have been emanating from the Grassland Biome study throughout its duration, a major terminal reporting and publication phase would begin in 1980" (pp. 9, 10).[22]

Apparently, Van Dyne had a different conception about the life of the biome program that he had not shared with his staff. I suspect that he envisioned the grassland biome becoming an ongoing, long-term study at Colorado State University, probably funded by a diversity of sources, and gradually becoming oriented to applying a tested and validated grassland model to regional problem solving. This was a perfectly reasonable and valid dream, and to a large degree it has come true because the Natural Resource Ecology Laboratory continues its grassland studies. Putting an ending date in the 1973 proposal that could be used to terminate the funding of the large centralized effort was, however, perhaps in Van Dyne's mind a self-defeating and disastrous strategy.

By taking sabbatical leave in 1972–73, Van Dyne lost firm control of the program. His private vision of the future of the grassland program was not reflected in the statement that the program would end at the end of the decade. By leaving a group of able senior scientists to carry on biome management, he allowed the development of individual research initiatives and the evolution of group camaraderie in program direction. The former central control was weakened fatally. His outright repudiation of the proposal of June 1973, his attempt to reassert control, and his reaction to his own failure to convince NSF to return control solely to him further weakened any respect for his authority.

Nineteen seventy-four was the crisis year. NSF made a special site visit led by Eloise Clark, which apparently went well. The overall modeling activity under George Innis and the various component programs under other senior staff were strong and were following interesting scientific questions. Nevertheless, the rancor between Van Dyne, supported by those who respected his leadership and shared his goals, and other senior staff who wanted independence and a more open, collective leadership persisted. In March, Van Dyne was asked by Colorado State University to resign as biome director. Jim Gibson, who had been in charge of administration, was asked to serve as the project director, and the senior staff became directors of their projects. David Coleman received the first independent research grant (the second in the history of the program) in 1975; William Hunt received a second in 1977; and since then the group has continued to be funded through a variety of sources.

Van Dyne moved to the range science department at Colorado State University where he began a new modeling program, still focused on the grassland. He also retained connections with the biome program, and he was carried on their reports as an investigator until 1976. In 1981 he died unexpectedly of a heart attack; he was forty-nine.

Evaluation

In evaluating the impact of the IBP grassland biome program on ecosystem studies, one finds that there were many difficulties. The material developed by the program is voluminous and scattered. Nevertheless, three syntheses have appeared. The first was edited by Innis (1978a) and reports on the development of the systems model, which was called ELM. The second was edited by French (1979) and reports on some of the comprehensive studies. The third, edited by Paul Risser and co-workers (1981), is concerned with the true prairie ecosystem. In addition, two other volumes appeared in the Cambridge University Press IBP synthesis series and present comparisons of grasslands internationally. These volumes provide a foundation for an evaluation of the program.

We can analyze the program in several ways. First, from an economic point of view, we might audit its productivity in terms of the literature and graduate degrees. These parameters could be linked to its intellectual product. Second, we might ask, Did the project succeed in doing what it proposed to do at its outset? Third, we might question whether the program contributed to our understanding of ecosystems? Did it enlarge the conception? I will briefly explore each of these paths of analysis.

The IBP grassland project, through 1976, received $16.3 million in funding, uncorrected for inflation (table 5.1). We have only cumulative publication records through 1974, since those publications prepared in 1975 and 1976 appeared one or two years later and, as far as I know, were never included in a list of IBP publications. The publications, taking the raw lists from the proposals and correcting for duplicates, indicate that the production of titles held reasonably steady over the life of the project. There were from fifty-seven to seventy-six titles published annually. A count of listed titles, however, exaggerates the production, since it includes abstracts of talks published in program editions of bulletins, book reviews, and other minor writing. Nevertheless, these lists are what reviewers in NSF were given by the research team, and it represents the material by which the project chose to be judged. Besides published titles, the project produced 98 theses or dissertations during this period and 293 technical reports. The senior, (doctorate or equivalent) staff available to direct the research, prepare articles, and direct students increased over the life of the project. In the most active phase, there were about twenty person-years of senior staff time allocated and paid for. Of course, this amount does not include any volunteer time given to the project.

These data provide a basis for an economic analysis of the grassland project. From 1968 to 1974, the average cost for each published title was nearly $43,000, and for each published title plus thesis title, over $33,000. A few years earlier (1965), at the Savannah River Ecology Laboratory (SREL), I determined

Year	Dollars $\$ \times 10^6$	Senior-Staff Person-years	Titles Published	Theses and Dissertations
1968	0.85	3.35	—	1
1969	1.80	6.00	59	3
1970	1.80	10.47	72	7
1971	1.80	14.25	60	29
1972	2.94	20.10	26	27
1973	3.09	21.55	57	21
1974	2.5*	23.38	71	10
1975	2.6*	20.00	—	—
1976	2.5*	19.25	—	—

* requested amount
— no data available

5.1 Resources and production of the IBP grassland biome project, based on grassland biome project proposals and reports (University of Georgia library archives)

that the cost of each published title for the laboratory was approximately $10,000. The SREL costs did not include the construction of a building and other facilities, the operation of a multistate network of scientists, nor a large computer program as required in the IBP. These comparisons suggest that the grassland program was not excessively costly for its size and organization. Further, the number of publications produced by senior scientists was three-and-a-half titles per person-year, which is a reasonably strong rate of publication, especially for a new program. Thus, an economic evaluation of the grassland project suggests that it was productive given the resources available to it.

An economic analysis of basic research is unsatisfactory for several reasons. It is exceedingly difficult to determine the inputs. For example, Van Dyne never received his full salary from the grassland biome when he was biome director. How does one calculate the twenty-four-hour-a-day attention he gave to the project in terms of person-years or dollars? It is also difficult to establish outputs and their value. Some research has immediate value because it creates new directions and new paradigms. Other research has lasting value because it is foundational, providing baseline data that will always be used. Yet other research builds our understanding of a topic and adds to our knowledge. These thoughts bring us to our second form of evaluation.

The statement of objectives in the original proposal for the grassland biome (December 1967) is complex. It acknowledges that the purpose of the IBP was to examine the biological basis of productivity in human welfare. It then states that the project would be concerned with primary and secondary productivity and how those levels are affected by man. The immediate goals were identified as an analysis of "energy flow, nutrient cycles, trophic structure, spatial patterns, interspecies relations and species diversity." To reach the ultimate goal, the project would have to be involved in the analysis of structure, function, and the interaction of these processes in grassland function. The statement concludes: "The focal point of this research is to improve our understanding of entire systems. Throughout this study, the whole system will be kept continuously in view. No matter how narrow or detailed some of the projects may be, their relation to the whole will be the dominant theme."[23]

Finally, after over a hundred pages of detailed discussion of the aims and goals of the study of the abiotic environment and three trophic levels, ecosystem modeling is presented as the synthetic tool of the project. Models would serve not only to synthesize data but to organize and guide the research.

In the second proposal (January 1969), the objectives were made more general with the addition of the following three questions: (1) What are the driving forces making grassland ecosystems operate? (2) What are the major and minor components of grassland ecosystems and what are the changes in magnitudes of these components over time? (3) What are the important families or groups of processes which cause the interaction, coupling, or linking of the components of the ecosystem one to another?[24]

Thus, one of the main objectives was the development of a grassland ecosystem model. A series of models were produced (see fig. 5.4) (Van Dyne, 1978). The first several models were limited in scope. For example, several developed by Patten (1972) and W. G. Cale (1975) were linear models. In 1971, however, the team began developing the ELM model. ELM was a total system model focusing on biomass dynamics, which would be representative or applicable to sites in the grassland. The model addressed four questions:

1. What is the effect on net or gross primary productivity as the result of the following perturbations: (*a*) variations in the level and type of herbivory, (*b*) variations in temperature and precipitation or applied water, and (*c*) the addition of nitrogen or phosphorus?

2. How is the carrying capacity of a grassland affected by these perturbations?

3. Are the results of an appropriately driven model run consistent with field data taken in the Grassland Biome Program, and if not, why?

4. What are the changes in the composition of the producers as a result of these perturbations? (Woodmansee, 1978, 270)

Woodmansee (1978) reviewed and critiqued the ELM model program. His analysis suggests that the model was relatively successful, especially as an integrating and communicating device. It was less successful in answering the questions posed by Innis. For example, prediction of the impact of nitrogen addition on primary production was not verified (Breymeyer and Van Dyne, 1980, 398). The model predicted 161 grams dry weight per square meter of peak live biomass, and 290 were observed in the field. Woodmansee felt that these failures may have been owing to using Liebig's law of the minimum in the model, by which he meant that in the model the impacts from an environmental factor were applied one at a time when actually a synergistic response was probably operating.

Yet, these investigators built a successful ecosystem model, which fit the field data in many aspects. It was a model of a point in space, with time as the varying factor; it was also structurally conservative, using Lindeman's trophic levels as the key structural feature; and it did not easily respond to questions. It seems that every question required its own model. The ELM model did not so much produce answers as it led to questions. The model has not been used in the management of livestock grazing, apparently, because of its complexity, and the fact that it used a code that has gradually become outdated.

Thus, the idea of whole-ecosystem modeling, which was certainly a widely accepted and attractive idea in 1968, as tested in the ELM model did not have the impact expected. This was due partly to the rapidly expanding field of model building—the ideas of 1968 became outmoded rapidly. Partly, it was also due to social factors operating in the research team. The replacement of Van Dyne and Innis's later move to Utah allowed the direction to shift to components of the model, with the result that construction of a series of component models began (see fig. 5.4). This type of model fit more closely the strategy of the research community. Finally, the team was not asking the ecosystem model ecosystem questions. Rather, all the questions addressed components—productivity, cycling, and species impacts. The theory of ecosystems was not tested. Rather, it was accepted and used to build models and direct research. This is made especially clear in the second synthesis book edited by French. In this volume, a successful effort is made to compare data across sites within the comprehensive program. For example, in a chapter authored by French, R. K. Steinhorst, and D. M. Swift (1979), trophic pyramids for each site were compared. It was concluded that these pyramids did not vary over time but were significantly different between sites, thus justifying the identification of different types of grasslands. Questions about the validity of trophic pyramids apparently were never asked.

Thus, there was a basic contradiction in the IBP grassland biome study. The goal was a study of the whole system, yet the scientific questions were couched in terms of the behavior of trophic levels and components. No one, including Van Dyne, seemed to see this problem. It was at least partly due to the development of models without a spatial component. The ecosystem can be conceived as a spatial object and its behavior understood mainly as change in space over time.

Yet, the program was successful at the component and process level. Success came from adapting to findings and the experience as they unfolded. Trouble came from the conflict between the unbending direction of Van Dyne, which he hoped would lead to something new and useful, and the adaptive strategy of the senior scientists who were responding to the answers they were getting in the field. Possibly, if Van Dyne had been allowed to carry the program through to completion on his terms, something different would have happened. As it was, ecosystem modeling and research shifted from whole-system studies to component studies, and these have been supported continually ever since.

Van Dyne's dream of a new way to express ecological insight, which would be directly practical, was never achieved. His own evaluation of this failure was focused on the agencies supporting the effort. According to him, NSF and others never had the vision, confidence, or capacity to support the test. Van Dyne never influenced enough of his fellow ecologists to create the necessary pressure on the agencies to meet his expectations. Indeed, the history of the grassland biome program and the synthesis volumes indicates that the scientific community had little stomach for tight organization to meet abstract goals and little imagination about what might be needed for management of the biosphere. Instead, they relied on the tried and true methods of individual investigators, meeting together, discussing their work, and writing articles. The academic ecologists' viewpoint prevailed.

The grassland biome project was only one of five efforts to carry out ecosystem studies in the IBP. How did these other projects face such problems?

THE OTHER BIOME PROGRAMS
Besides the grasslands, biome projects were organized in the tundra, deserts, coniferous forests, and deciduous forests. An unsuccessful effort was made to

launch a tropical forest biome study. Each biome was organized and developed individually. The characteristics of the biomes reflected their leadership, organization, the environmental features of the biome, funding availability, and the state of prebiome knowledge. For this reason, it is difficult to compare one project with another. Even so, the criteria used in evaluating the grassland biome will be employed for other biomes. These criteria are the effectiveness of the organizational structure, project productivity, and relevance to ecosystem studies.

No biome program other than the grassland biome program followed the original plan of a central site, with satellite comparative sites. This form of a spatial context was lost, partly because of the cost and effort to organize satellite programs—the central programs were never supported at the level requested—and partly owing to a change in program conception. The grassland model was to be an abstract general model that could be modified with data from other sites. In this way specific site-controlled problems could be addressed with a universal ecosystem model. The other biome programs tended to copy the grassland program, and when questions about central modeling were raised, attempts to construct a single-biome model were abandoned.

Rather, each biome discovered that local, site-controlled differences were a crucial elements of its system and needed to be considered in any general scheme. The tundra biome program focused its work at Point Barrow, Alaska. The work of the desert biome program was carried out at several sites, with the one at Curlew valley, Utah, assuming something of a central location. The coniferous forest biome program had two sites, the Cedar River basin near Seattle and H. J. Andrews experimental forest in Oregon. The deciduous forest biome program had five sites. In each of these biome programs, the number of sites reflected a compromise between the inclusion of ongoing ecosystem programs in the biome, a need to create a central tendency of the biome for theoretical modeling purposes, the past experience of ecologists, and the existence of banks of information.

The biome programs also deviated from the central organization of the grassland program. There were no other leaders with the personal characteristics of Van Dyne. The tundra program was led by Jerry Brown, a soil scientist with the U.S. Army Cold Regions Research and Engineering Laboratory at Hanover, New Hampshire. This biome program relied on the remoteness of the site, the compression of research into a few summer months, and the experience of working together in the field to organize research. The pattern was fundamentally no different from former arctic studies at Point Barrow. The desert biome program was led by David Goodall, a systems modeler formerly from Australia and before the IBP, a professor at the University of California at

Irvine, and by Fred Wagner, a professor of wildlife biology at Utah State University. Goodall did have strong opinions about modeling and tried to control the centripetal tendency of the biome program but the diversity of the desert environment and the individualism of the researchers defeated him. The coniferous forest program was led by Stanley Gessel, a professor at the University of Washington School of Forestry. This biome program incorporated scientists from the School of Forestry and the U.S. Forest Service who were used to organized and directed research; it did not suffer from organizational problems. The biome research was located at two centers that functioned almost independently. Finally, the eastern deciduous forest biome program was led by Auerbach, the manager of the environmental science program at Oak Ridge National Laboratory. In Auerbach, the IBP probably had the most experienced manager of ecology work in the United States. He had organized and guided the development of his program into one of the foremost in the world. For Auerbach, the problem was that he had numerous well-organized and active ecosystem programs in the biome. Only a few of these were willing to be part of the project, and they exercised an independence that was impossible to curtail.

The consequence of this variety was that the Analysis of Ecosystem program operated on expediency, taking advantage of the earlier development of ecosystem studies and allowing the scientific community to organize itself to meet its own goals and seek its own advantage. Given the strong opinion of individualist ecologists that biome organizations were fundamentally inefficient, supported mediocre science, and were politically suspect, it is obvious that the organizers of the Analysis of Ecosystems program made the appropriate decision. Yet, the compromise meant that ecology lost its opportunity to organize hierarchically and use funds to benefit the entire community. The argument that there was a right level of organization for ecological research and that those studying ecology at other levels were misguided at best, or intellectually dishonest at worst, was ultimately destructive to ecology as a science and prevented it from achieving its potential in the decade of the environment (1965–75).

The Analysis of Ecosystem program began in 1967 with the intention of applying a systems approach to ecological studies. The systems approach was intended to organize research, as well as to produce predictive models that could link the theoretical studies of ecosystems to applied environmental and resource problems. Systems ecology did advance with the IBP projects, partly as a result of focused and increased support, attraction of modelers and mathematicians such Goodall, Innis, and others to the ecosystem concept and the interaction of field scientists, modelers and programmers, and applied researchers. Their interaction produced a tremendous amount of argument but also a check on a too strong reductionism and too abstract modeling. A major argument that was not adequately resolved during IBP was the degree of disaggregation of the ecosystem needed for accurate models of real-world systems. The trophic-level concept was the disaggregation theory of choice, with the levels being decomposed into groups of like-acting species populations. This level of decomposition satisfied neither the community ecologist, who was concerned with interactions between single-species populations, nor the evolutionary ecologist, who argued that selection occurred at the level of the individual organism and that higher levels were abstractions. Regardless of the validity or lack of validity of these opinions, it was not possible to represent many populations in a ecosystem model, let alone individuals, and stay within the capacity of the computers and the imagination of the modelers. Scott Overton, of the Coniferous Forest Biome program at Oregon State University, proposed that models be constructed of subsystems and that the linkages between subsystems be made clear and realistic. With this structure, one could attend to those subsystems of interest and their linkages and ignore others. Overton's idea of disaggregation was never operationalized during the IBP, except in his models of the coniferous forest.

Systems models were most useful in organizing research and showing researchers how systems were constructed. The development of predictive models was less successful. The grassland program produced the ELM model, which was a total grassland model, and the tundra biome produced word models of the tundra. The tundra word model described the tundra ecosystem in a few paragraphs. Although both of these were effective within limits, the ELM model was not sufficiently adaptable, being strongly controlled by its structure and formatting. The other programs created practical models that focused on processes, such as nitrogen cycling, or on components, such as decomposer populations in the soil. These models proved highly effective, and the growth of ecosystem modeling is derived from their success. Every biome program produced useful models at levels of scale below the entire ecosystem.

The Analysis of Ecosystem program had a large impact on ecosystem studies, independent of the science of ecology, which was mainly through the institutionalization of ecosystem studies in the United States. First, funding, which amounted to about $57 million over the life of the program (Fred Smith had predicted the cost would be $45 million) was transferred intact to a new program in the division of environmental biology of NSF. The program was called "Ecosystem Studies," and it continued to provide funding for ecosystem work after IBP ended. The consequence of this action was that the new source of support garnered by IBP was kept for ecological science and the partly completed biome projects could be ended by synthesis activity over several years rather than abruptly terminated and lost. All the biomes produced synthesis reports of some type as a consequence.

Second, the biome program also provided support for a new set of academic centers and opened the possibilities for ecosystem studies directly to the universities beyond the earlier connection through AEC national laboratories. These new centers included the University of Georgia, Colorado State University, Utah State University, San Diego State University, and Oregon State University. These centers expanded ecological research across the United States, opening many new opportunities for ecologists of all types and linking theoretical ecologists with applied ecologists. In most cases, these programs have become institutionalized within universities and continue activity today.

Finally, the biome programs involved over 1,800 scientists. Many of these people became familiar with ecosystem studies through the IBP and then they continued this type of work afterward. The time of the IBP coincided with development of environmental management programs in the states, the federal government, and in the private sector of the economy, and many of those trained in the biome programs found employment in environmental management and protection. Even though departments of ecosystem studies or graduate programs in ecosystem studies did not become common, the value of this type of training became widely recognized and many students were encouraged to do theses on the processes or components of ecosystems.

The impact of the IBP biome program on the development of ecosystem theory was marginal. One might say that the IBP carried bricolage to a new level of activity and scale. There were enough theories or concepts in the literature so that almost any project could be underpinned theoretically. The programs were not designed to sort out competing or contradictory ideas. Rather, they were driven, at least initially, by the idea that ecologists could construct a mechanical systems model built on the concepts of trophic levels, the food web, or the food cycle, and then represent the dynamic behavior of the components by data from organisms or populations that are surrogates of the component. This "bottom-up" or "design-up" approach did not prove possible or useful. Further, the biome projects did not effectively promote landscape ecology, as Odum has hoped. The biome was the setting for site research but was not really addressed as such in an effective manner.

Did IBP contribute to ecosystem studies? Yes, definitely. Did it achieve the objectives set at its start? No, not entirely. Did the field advance theoretically? No, not significantly. Did it build the institutional, manpower, and structural base for further advances? Yes. Did it further ecological knowledge? Yes, decidedly. The corpus of literature produced by the IBP stands as testimony to its contribution to all aspects of ecological science. Finally, Did it contribute to the solving of ecological problems? Yes, in many cases it did so directly. But possibly it is the indirect contributions it made in creating practitioners, methods, and information that was even more impressive.

CHAPTER 6

Consolidation and Extension of the Concept

The U.S. biome program dominated, and in many peoples minds characterized, the IBP. Yet the biome program continued the mixed research approach
that had characterized ecosystem studies in the post–second world war period.
The movement in the IBP period was so rapid, so many investigators were
involved, and their motivations so different that the little coherence was obtained even within single biome programs.

Actually, the biome program created an opportunity for ecosystem studies
to break out of a confused situation, an opportunity to test approaches to
studying ecological systems. The test was conducted unknowingly within the
IBP, but it was through a comparison of the paradigmatic IBP biome project,
the grassland biome project, and another program outside the IBP biome that
made this test visible and understandable. This other project was called the
Hubbard Brook project since it was located at Hubbard Brook Experimental
Forest.

The test involved approaching ecosystem studies from the components,
which would then be linked together into a system in a computer model-driven
theory or as a natural object that could be studied using conventional scientific
methods. In this latter case, the ecosystem object would be observed, a pattern
of behavior established, and questions about the origin of this behavior posed.
These questions would then require that the ecosystem be dissected into
components or subsystems and their linkages and their behaviors observed and

explained. In this way, a mechanistic explanation of ecosystem behavior could be framed in terms of biological, chemical, or physical principles.

There were such studies within the IBP biome, and the contrast between the approaches engendered some of the disagreement that characterized the IBP. The distinction was never institutionalized or made a major element of debate. The grassland biome most closely followed the first, bottom-up, approach to ecosystem studies. The desert biome program also took this approach but could not carry it as far as Van Dyne had pushed the grassland project. The tundra project began with the intention of following the approach but quickly abandoned the effort and focused efforts on terrestrial and aquatic sites, using conventional techniques. The coniferous forest biome program had two sites: one at the Cedar River basin was organized in the bottom-up mode, while the other at H. J. Andrews took a top-down approach. Finally, the Eastern Deciduous Forest Biome project (EDFB) was largely organized as top-down ecosystem studies.

What seems most important in fashioning the style of research was the presence or absence of easily recognized boundaries of the ecosystem. Where projects focused on watersheds—including those at H. J. Andrews, Coweeta in North Carolina, Madison, Wisconsin, and so on—the ecosystem as an object focus was obvious and was adopted naturally. Where ecosystem boundaries were less clear, even when the watershed approach was adopted, as at the Pawnee site in Colorado, it was easier to adopt the modeling approach and to construct an ecosystem by summing its parts into a whole system.

The comparison of approaches was not made explicit within the IBP biome program. Rather, it became apparent through an evaluation of the IBP biome by Battelle Memorial Institute, Columbus, Ohio. Immediately as IBP ended in 1974, NSF was faced with moving the large quantity of monies funding the IBP into new or ongoing programs. To do so intelligently it needed a review of IBP, even though the project had just ended officially and was actually in the midst of a synthesis phase. Actually, it would require five to six more years before the bulk of the synthesis treatises and books would be published. Battelle was chosen to make the evaluation, which took the form of a comparison of three biome projects: the grassland, tundra, and the EDFB, with another large ecosystem-oriented program. Battelle chose to compare the biomes with the Hubbard Brook effort. Hubbard Brook was a watershed study and it used a conventional scientific method to study the watershed ecosystems. Further, its leaders and organizers, Bormann and Likens, understood their approach and promoted and defended it in their proposals, speeches, and articles. The Hubbard Brook project became the key to establishing the future direction for ecosystem studies.

HUBBARD BROOK

In 1962 ecologists Herbert Bormann and Gene Likens (figs. 6.1 and 6.2), began to organize the Hubbard Brook study that would have such a significant impact on the development of ecosystem studies. Bormann, a plant ecologist, had been a student of H. J. Oosting at Duke University and was interested in the function of trees and forests. He had become acquainted with the hydrologic studies at the Coweeta Hydrologic Laboratory near Franklin, North Carolina, while he was a graduate student, and he took his own students there when he was a professor at Emory University in Atlanta (Bormann, 1985). Bormann moved to Dartmouth University in Hanover, New Hampshire, in 1956, where he was introduced to the U.S. Forest Service's hydrological facilities at Hubbard Brook Experimental Forest. In 1960, he had the inspiration to study mineral cycling on small watersheds. His concepts were expressed in a letter to Robert Pierce of the U.S. Forest Service Research Center at Laconia, New Hampshire, in November 1960:

> The other day while discussing the problem of mineral cycling through ecosystems, the thought came to me that your installation at Hubbard Brook represents a veritable research gold mine in regard to fundamental studies on mineral cycling.
>
> One of your small watersheds with a weir at the outlet represents a perfect area for controlled research. If one were to select one or several minerals, such as K+, it would be possible, by taking weekly water samples and analyzing them, to determine quantitatively the amount of K+ leaving the system. Weekly estimates of K+ per liter multiplied by the liters of water leaving the watershed would give the quantitative figure. Since the watershed is theoretically tight and all water falling on the shed appears at the weir (excepting evaporation which would not remove any minerals), the quantitative figure would represent total loss of K+ from the watershed (excepting leaf litter blown out, or presumably this would be counterbalanced by leaf litter blown in. The same argument goes for other losses and additions due to animals, etc.).
>
> Some minerals may be added by rain or snowfall, therefore both rain and snow would have to be analyzed for the mineral(s) in question. These analyses multiplied by the amount of rain or snow would give the total amount of the mineral(s) added to the system.
>
> By subtracting the total amount added from the total amount lost, it would be possible to estimate the steady-state losses from the system. Theoretically the only place these minerals could come from is the underlying parent material and bedrock. Thus, the loss represents the rate

6.1 F. Herbert Bormann. Photograph courtesy of F. H. Bormann, 1991

at which the bedrock is wasting away in terms of the mineral(s) under consideration. By knowing the chemical composition of the bedrock, it would be possible to determine the rate at which it is breaking down.

This figure would seem to be of considerable consequence because it would quantify the rate of erosion, it would shed considerable light on the rate of soil formation, and it would also tell something about the rate at which minerals useful to plant growth are added to and lost from the system. The latter might lead into further studies of how various treatments affect the mineral cycling patterns.

Further implementation of Bormann's ideas came in 1961 when Likens joined the faculty at Dartmouth temporarily, returning from Wisconsin the next year as a regular faculty member. Likens is a limnologist who studied with Arthur Hasler at the University of Wisconsin where he became familiar with Hasler's view of the integration of aquatic and terrestrial systems into one landscape unit. Thus, Likens came to Dartmouth with a similar theoretical

6.2 Gene E. Likens. Photograph courtesy of G. E. Likens, 1991

understanding as Bormann but with special competence in the study of aquatic systems. These two developed the theoretical and practical aspects of a small watershed approach to biogeochemical cycling, which was later published as an article in *Science* (Bormann and Likens, 1967). Likens also began studying Mirror Lake, a small lake in the Hubbard Brook watershed. The following year, in 1962, Bormann and Likens were joined by Noye Johnson, a geochemist on the geology department staff at Dartmouth; in 1963, together they wrote the first proposal to NSF for an ecosystem project at Hubbard Brook. In the early years of the project, the three were associated with Robert Pierce, a forest hydrologist of the U.S. Forest Service, and John Eaton, a forest ecologist.[1]

The Hubbard Brook study blazed several new paths in ecosystem research. It replaced energy flow with nutrient cycling as the principal functional process in the ecosystem. It focused research on the discrete, easily recognized hydrologic systems called watersheds. Where before ecosystem boundaries tended to be arbitrary, at Hubbard Brook the hydrologic divide defined the limits of the system in a natural way. Further, the focus on chemical nutrients created links with geochemistry, soil science, hydrology, and atmospheric sciences, which were more akin to ecological interests than were theoretical physics and information theory, with their arguments on the ecological meaning of entropy and feedback. In a sense, the Hubbard Brook project brought ecosystem studies back to earth.

Hubbard Brook is located in the White Mountains of north central New Hampshire. The climate is humid continental, with short, cool summers and long, cold winters. The experimental forest covers about three thousand hectares and ranges in altitude up to 1,015 meters. The vegetation of this forest is northern mixed hardwoods and conifers, with sugar maple (*Acer saccharum*), American beech (*Fagus grandifolia*), and yellow Birch (*Betula alleghaniensis*) as the characteristic species. The forest was cut in 1910–19 but has been undisturbed since (Likens et al. 1977, 5).

The nutrient cycling approach to ecosystem studies was announced in an article in *Science* in 1967, the year the grassland biome program was started. In the article, Bormann and Likens (1967, 424) argued that small watersheds are ideal sites for ecosystem studies because they can be used to measure weathering and erosion, the hydrologic cycle, and the movement of chemical elements between biological components. As they suggested, the problem was that ecosystem studies heretofore had focused on those aspects of biogeochemical cycling that occurred within ecosystems. They were proposing to consider the ecosystem as an object and to expand these studies by linking them with the climatic processes of the environment and to other systems. They stressed the physical-chemical interactions with the hydrosphere and lithosphere, under the assumption that "the rate of release of nutrients from minerals by weathering, the addition of nutrients by erosion, and the loss of nutrients by erosion are three primary determinants of structure and function in terrestrial ecosystems."

Bormann and Likens also presented a new conceptual model of the ecosystem in their article (fig. 6.3), in which the ecosystem is linked to the biosphere, and organic and inorganic components are connected. They recognized four compartments: (1) the atmosphere, (2) the pool of available nutrients in the soil, (3) the quantity available in living and nonliving organic materials, and (4) the soil and rock minerals. These compartments are emphasized in their conceptual model.

A key point of their article was that ecosystem processes are closely coupled to the hydrologic cycle. The nutrient budget for a single element in a watershed ecosystem was given as: meteorologic input + biologic input − geologic output + biologic output = net system loss or gain. Nutrient outputs are related to hydrologic parameters, such as variations in stream flow, precipita-

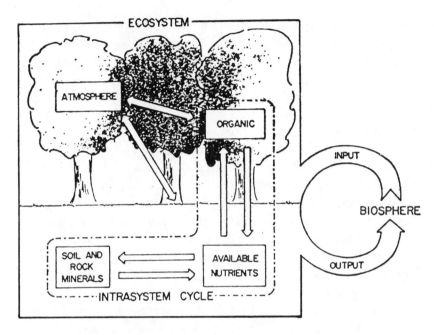

6.3 Nutrient cycling diagram for Hubbard Brook, showing sites of accumulation and major pathways (Bormann and Likens, 1967)

tion, and evapotranspiration, and to phenologic events. The weathering of minerals can be estimated from net losses of the element from the watershed. Further, the method allows for a comparison of the dissolved elements in solution in streams and those in the solid material in the bed load. Finally, Bormann and Likens pointed out how an experimental approach to watershed studies could be developed by treating the systems in various ways, and they called for a network of watershed ecosystem studies throughout the United States.

In 1977, Likens and Bormann, with Pierce, Eaton, and Johnson, published the first book-length report on Hubbard Brook, which summarized a large number of technical monographs and reports.[2] They presented a series of significant findings on the biogeochemistry of ecosystems. First, they showed that the forest ecosystem acts as a chemical buffer or filter for many elements, including pollutants that enter the system through the atmosphere. These pollutants may be stored and fixed within the system or their release to the environment may be slowed down. Second, the undisturbed system has predictable outputs in stream water. Water output is tightly coupled to meteorologic inputs, and the annual variation in output tracks the variation in input. Since meteorological inputs vary, long time sequences are required to understand the cycling of each chemical element. Third, the forest both gains and loses elements. In an absolute sense, the forest is gaining nitrogen, sulfur, phosphorus, and chlorine and losing silicon, calcium, sodium, aluminum, magnesium, and potassium. Weathering is the main reason for losing elements; and the system is susceptible to disturbance by external influence. For example, it had received sulfur from industry through the atmosphere, even though Hubbard Brook is 100 kilometers from any concentration of industrial activity, and this input had had an impact on the nutrient cycles.

The Hubbard Brook approach to the ecosystem as an object of scientific study led to numerous causal questions that became the focus of research by several generations of graduate students, as well as visiting and associate investigators. As a consequence, Bormann and Likens were able to explain how ecosystems functioned in their summary articles and books. An example of this type of detailed investigation is Judy Meyer's examination of phosphorus in the streams draining Hubbard Brook watershed. Meyer was a student of Likens at Cornell University.

Meyer (Meyer and Likens, 1979) used a conventional ecosystem analysis to contrast the input and output of phosphorus in a headwater stream draining a forested watershed. Phosphorus was partitioned into three fractions: dissolved, fine, and coarse particulate. The inputs of the three fractions were almost the same, but fine particulate matter in the stream dominated the output. Meyer paid special attention to the temporal changes in phosphorus dynamics. Because the particulate concentration increased exponentially with stream discharge, export was associated with high discharge events. At these times only 9 percent of the water but over 50 percent of the annual phosphorus release left the system. Meyer commented that "the forest, by controlling the hydrology, controls the P balance in the stream."

Another strength of the Hubbard Brook approach to ecosystem study was that it allowed investigators to set Hubbard Brook in the landscapes of the northeastern United States. This made it possible to reason from one site to a landscape and made the results of studies at one site of greater practical value. Bormann and Likens did not make this landscape linkage explicit, nor did they place Hubbard Brook in a landscape hierarchy. Yet these steps came easily once the ecosystem was coupled to geologic, atmospheric, and other ecological systems.

Extension of the results of the Hubbard Brook study to the wider landscape and to environmental issues in that regional landscape also led to controversy and personal attacks on the principal investigators. The first controversy concerned the impact of forest cutting on stream chemistry. Bormann and Likens reported that the nitrogen levels in stream water increased after the forest was cut (Likens et al., 1970; Bormann et al., 1974). This is hardly a surprising finding to an ecologist, because the biotic compartments that regulate nitrogen cycling are radically disturbed when trees are removed from a forest and the temperature and other environmental conditions at the ground level are substantially changed, yet foresters reacted with alarm. The response of professional foresters paralleled that of agriculturists when Rachel Carson condemned thoughtless pesticide use in 1962. Questions were raised about the accuracy of the Hubbard Brook data, and the researchers' motivations were called into question. Nevertheless, the study was sound, the data were convincing and easily corroborated, and the logic of nutrient cycling was clear. Forest clear cutting with heavy machinery can be extremely disturbing to the delicate and subtle processes of forests, and only a person focused exclusively on economic or political criteria would find this surprising.

A second controversy stimulated by the Hubbard Brook data demonstrated the broader application of the data to the landscapes of the Northeast. This controversy was based on the first report of acid rain effects on a forest ecosystem in the United States (Likens and Bormann, 1974; Likens et al., 1976; Likens et al. 1979). The impact of acid deposition had been reported in Scandinavia before this time (Rodhe, 1972; Mysterud, 1971), but people assumed that the disturbance was restricted to a small part of Europe where concentrated industrial activity was present. Likens and Bormann showed that these problems were much more widespread. Air pollution generated outside of the northeastern region was affecting the forest ecosystems in the mountains. Again, the reaction to these findings was an attack on the data and the scientists, not on use of the data and findings in enlightened decision making aimed at increasing public health and well-being. The media made a search for the culprits producing sulfur and nitrogen atmospheric pollution. Blame was placed on the Ohio Valley, New York, Pennsylvania, New Jersey, and Canadian industries and on the urban concentrations of the New York megalopolis. Industrial and political forces reacted by questioning the results and calling for yet more studies or arguing for jobs against the environment. Few in government or decision-making positions seemed able to extrapolate from the problems of the Northeast to the biomes, the eastern region, or to the nation. Likens and Bormann were bitterly attacked. Yet the data were consistent and were supported by other studies. The message from the Hubbard Brook studies was sound. Air pollution is a problem in the Northeast and elsewhere in the United States, and we know the sources of this pollution and many of its consequences.

The Hubbard Brook project was also controversial within the science of ecology. Bormann and Likens had strong personal feelings and beliefs about science management and responsibility that led them into conflict with other ecologists, especially with reviewers of their proposals for funding the Hubbard Brook studies. Their difference in approach first became apparent when they did not respond to an informal invitation to participate in the Eastern Deciduous Forest Biome project. IBP was clearly developing on a path unlike that of Hubbard Brook, which was already under way on a well-designed research plan. In retrospect, Bormann and Likens refusal to join the IBP biome study was a significant gain for ecological science. Their research approach proved to be scientifically sound. By applying the conventional methods of science to the ecosystem object, it provided the experience that moved ecosystem studies beyond the analogical, metaphorical stage of the mid-1960s to another level.

In the mid-1960s, when the IBP biome programs were being organized, these theoretical distinctions were not clear. Rather, the estrangement between the project at Hubbard Brook and the biomes hinged more on matters of organization and direction. Bormann and Likens were forging a particular approach to ecosystem studies and were unwilling to subordinate themselves to another organization that was operating on other principles. They prided themselves on allowing investigators a strong measure of independence. Further, Hubbard Brook was managed by a directorate. Bormann (1985, 3) commented on the organization: "We met frequently, reviewed progress and discussed future directions. The necessity to write frequent grant proposals, although burdensome, provided the basis for intensive evaluation of research directions and for concluding or adding new research. In my opinion, this harmonious relationship more than any other single thing is responsible for the longevity of the Hubbard Brook Ecosystem Study. No one of us could possibly have done it alone." In addition, Bormann and Likens required each investigator to obey the Forest Service rules regarding use of the site and made the sharing of data mandatory. Further, in the early days of the project each investigator was asked to list the potential documents to come from a proposed study and to list the authors in probable order. These issues were considered so important that in Bormann and Likens's first book the only comment italicized in the preface was one addressing the individuality of research design and execution: "We deem this individual research freedom one of the greatest assets of the Hubbard Brook study" (Likens et al., 1977, vi).

By stressing their theory of research management to the reviewers of their research proposals and avoiding participation in other large-scale projects, Bormann and Likens caused themselves a great deal of trouble. After all, most reviewers of ecosystem proposals were people who had made commitments to the biome type of project, and almost every Hubbard Brook proposal earned some negative marks from reviewers for their method of management. The quality of the data and the productivity of the Hubbard Brook project always saved the proposal, but Bormann and Likens suffered greatly for their position and the way they expressed it.

Actually, the management of the Hubbard Brook project was more complex than Bormann and Likens have reported in proposals or published reports. While the team of senior investigators led by them was a key element, their way of managing incorporated a discussion of issues and alternatives by a diverse group of experts with expertise and information in a variety of fields. As Likens and Bormann became known internationally through the Hubbard Brook project, their influence became more dominant, but possibly more subtle. Hubbard Brook was and is an academic project; it is dominated by collegial relations between professors and by a professor-student relationship, not the manager-employee relationship of an institute or laboratory. Not least important in creating a sense of teamwork, as Bormann pointed out, was that they lived at Pleasant View Farm, where students and faculty shared a research life together. These factors created a special feeling of being part of something important.

The evaluation of the IBP by Battelle (Battelle Columbus Laboratories, 1975, I–41) emphasized these differences between Hubbard Brook and the various biome programs. The evaluation stated that Hubbard Brook investigators had focused more strongly on abiotic factors and transfers from the abiotic to the biotic compartments than had the other biome programs: that is, Hubbard Brook was a study of biogeochemistry. It also developed more slowly, involved fewer people, and cost less money (table 6.1). The reviewers reported that the ratio of dollars for each publication for Hubbard Brook was $13,000, while it was $31,000 for the tundra project, $47,000 for EDFB, and $59,000 for the grassland program. The reviewers commented: "The biome programs did not undergo a gradual evolution and development as did Hubbard Brook but, instead, exploded on the research scene powered by massive funding, ambitious goals, and short deadlines. Difficulties and inefficiencies were inevitable." The Hubbard Brook project served as a model ecosystem project for many investigators, and it continues this role today, not only as an integrated ecosystem project, but also as a part of the long-term ecosystem research program of NSF. This means that the more conventionally scientific approach at Hubbard Brook has become the norm of ecosystem projects rather than the exception.

IBP OUTSIDE THE BIOME PROJECTS

While Hubbard Brook created an alternative to the IBP biome studies within the United States, the international production terrestrial projects of the IBP

Year	Program			
	Deciduous Forest	Grassland	Tundra	Hubbard Brook
1969	200	851	0	142
1970	229	1,800	351	182
1971	1,200	1,874	1,003	178
1972	1,992	2,055	1,192	193
1973	2,100	1,997	998	192
1974	1,800	1,800	936	200
Total	7,521 ·	10,377	4,480	1,087

6.1 A comparison of the costs of three IBP biome projects and the Hubbard Brook project from 1969 to 1974 (in thousands of dollars) (Battelle Columbus Laboratories, 1975)

continued on their planned path, contributing fundamental knowledge about the ecology of species and groups of species. Nevertheless, the apparent success of the biome projects led ecologists in other countries who were not involved in the PT programs to propose biomelike projects too. A variety of these were carried out in northern Europe and Japan. One of the most successful was the Solling project in the Federal Republic of Germany organized by Heinz Ellenberg of Gottingen University. Ellenberg had been involved in the early IBP conference with the IUCN and was a key organizer of terrestrial work in Europe. He was able to gain support for an IBP pilot project in Germany.

The Solling Project

Ellenberg, a plant ecologist with wide experience and training, had a strong reputation in German ecology and brought a different perspective to ecosystem studies than did the Americans. As a young man he had been an assistant to Reinhold Tuxen, a vegetation scientist who was a major proponent of the Braun-Blanquet school, and Heinrich Walter, an ecological plant geographer known for his world climate atlas and description of the vegetation patterns of the earth.[3] Ellenberg's interests extended from this classical background in vegetation science and biogeography to many other topics in plant and general ecology. He used the ecosystem concept as an organizing device to link different kinds of scientists in a single project. The German IBP committee decided to concentrate its efforts in a pilot project under Ellenberg's direction that, if successful and interesting, could later be expanded and replicated. This pilot study was located in a low mountainous region (400–500 meters elevation) near Gottingen. The area, called Solling, contains a mosaic of communities from seminatural forests to agricultural fields.

One of the features of the Solling project that was of interest in the context of the U.S. biome program, was that it took a landscape approach from the beginning. Indeed, in the final report, written in 1986, Ellenberg began with the question, What is the ecosystem? *(Was ist ein Ökosystem?)*, answering it with a description of a gradient of systems ranging from naturallike *(naturliche)* ecosystems, nature-resembling ecosystems, transformed ecosystems, and disturbed or degraded systems (Ellenberg, Mayer, and Schauermann, 1986, 19). In the Solling area research was directed toward examples of these types of ecosystems including the acidophilous beech forest and planted spruce forest. It considered forests of both types at several different ages, along with permanent grassland and cultivated fields that were in annual grass. This was an ambitious program that took a geographical, landscape approach, while concentrating research in a single place.

When work on the Solling project began, the scientists had the advantage of beginning work with a comparatively good knowledge of the flora, fauna, climate, soils and general ecology of the communities. Although Ellenberg cited gaps in knowledge and areas where it was not possible to go deeply into the analysis, the background for an ecosystem study was probably better there than in most places where IBP work was undertaken. Studies went forward in all the appropriate areas from a description of the abundance and productivity of vegetation, animals, and microorganisms, to plant physiology, nutrient cycling, energetics, climate analysis and soil chemistry. Because this was a pilot study and also one of the first IBP projects to get under way, its methods were described in an English-language book published in 1971 by Springer-Verlag (Ellenberg, 1971). The final summary report, in German, was published in 1986 (Ellenberg, Mayer, and Schauermann, 1986).

The Solling project was an ambitious effort involving many scientists from diverse disciplines and resulted in hundreds of individual articles and reports in the literature. The summary book provides a good abstract of the effort, since the many chapters were authored by the individual scientists involved. Unlike the biome programs of the United States, there was no attempt to force the results into a single synthesis through a model or abstract theoretical device. Instead, each part was placed within a landscape and a conceptual ecosystem model and developed a theme within its own logic and tradition. A major emphasis was placed on vegetation, but this emphasis does not seem unbalanced when compared to other studies.

The Solling project was a strong ecological project. It does not appear biased toward the biotic community to the neglect of the abiotic elements in the environment, nor does it deal too much with description to the neglect of function. Although it delivers few surprises, in the context of its purpose and design, the Solling project was exceptionally successful. It was a pilot project that provided sound data on ecosystem structure and function, thus building the scientific conception of the ecosystem. Further, it provided a basis from which to reason about the causes of forest dieback, which became a serious problem in the Federal Republic of Germany in the 1980s. Nevertheless, when it was completed, the project did not lead to establishment of other ecosystem centers nor to institutionalization of ecosystem research in Germany. This failure was partly because of the lack of enthusiastic leadership (Ellenberg retired soon after) and partly owing to the disinterest of German ecologists in ecosystem studies. In the 1980s, a European network of catchments organized for ecosystem research was started, which includes several German sites. This network is called ENCORE.

By the time IBP ended many countries had carried out process studies that contributed to ecosystem science. Nonetheless, most of this work focused on natural ecosystems within the boundaries of technologically advanced countries, which could put together the teams and equipment for full scale scientific research. This was one reason for Germany's placing its resources in a single pilot project. Most IBP national programs were only minimal, focusing on a single question or a few questions. Nevertheless, even this minimal effort was too much for many developing countries and few studies were carried out in tropical or arid-zone countries. Among the exceptions were the research programs organized by the French in the Ivory Coast, West Africa, the study of the Pasoh rain forest in Malaysia by scientists from Malaysia, Britain, and Japan,[4] and the study of Chakia forest in India by ecologists at Benaras Hindu University, Varanasi. Ecologists had started a large-scale investigation of the savanna in the Ivory Coast in 1961, and this program became incorporated into the IBP. I discuss it below as an example.

The Lamto Project

In 1961, French ecologists associated with the l'Ecole Normale Superieure de Paris, supported by the Centre National de la Recherché Scientifique (CNRS), organized a program with the University of Abidjan to study the ecology of the savanna on the edge of the Sahel in an area called Lamto. This savanna is called the Ronier palm savanna, since Ronier palms (*Borassus aethiopum*) rise above the grassland and create the characteristic aspect of its vegetation. The savanna burns regularly (Lamotte, 1969). It occurs on the uplands and interacts with dense gallery forest along stream margins. The interdigitation of these several types of communities creates a complex landscape.

The work at Lamto proceeded in the usual pattern, first with a reconnaissance of the flora and fauna and measurements of climatic factors, followed by studies of trophic exchange. The trophic studies were organized by individual populations of organisms and focused on energy flow, productivity, and respiration. A great deal of attention was paid to aboveground and soil animals, because the work originated in the laboratory of zoology. Also, Maxime Lamotte, professor at L'Ecole Normale Superieure de Paris and director of the project, attracted zoologists and ecologists from other organizations and countries to Lamto. There was no attempt to create system models of the savanna or to experiment with different modes of social organization of scientific research.

The Lamto project was one of six within an ambitious comparative IBP program by French ecologists. Besides Lamto, the other ecosystem studies were on permanent pastures in northwest France, the forest at Fountainebleau near Paris, the Mediterranean oak forest near Montpellier, the Sahel savanna of Senegal, and the tropical rain forest of the Ivory Coast. These studies were synthesized in a book edited by Lamotte and François Bourliere (1978), who was president of the Scientific Committee on the IBP of ICSU. The synthesis is unusual because of the strong connection of its studies with the ongoing currents in population and community ecology. For example, in the introductory chapter, these zoologists discussed coevolution, niche theory, and competitive interactions that structure the populations within the ecosystem and provide the basis for trophic webs and chains, which were the central focus of ecosystem ecologists. These topics almost never received attention in the U.S. biome projects.

The Lamto study also focused on the biology of species. Although there was an intensive study of vegetation, with measurements of its productivity in space and time as well as before and after fire, the major attention was on animals. A special sampling technique was employed for many of the animal studies where a large number of fieldworkers surrounded a plot and literally picked it apart by hand. The availability of sufficient field labor was a key to obtaining sound censuses of animal populations. The larger herbivores, characteristic of African savannas, were observed infrequently in this locality. Rather, herbivory was from smaller mammals, other vertebrates, and insects. The population censuses provided the basis for study of the energy flow of the populations using conventional respirometry and productivity approaches, with all the problems of extrapolation and measurement.

Lamotte (1978, 294) used an energy flow diagram to synthesize the results of the Lamto study and to show the relative significance of groups of taxa. The net primary production of the Lamto vegetation was primarily derived from the grasses and herbaceous species and amounted to 114×10^6 kcal/ha/year. Of this amount, fire consumed 36 units, herbivores consumed 62 units, earthworms accounted for 55 units, and termites 4 units. Rodents and insects were relatively insignificant. Sixteen units were unconsumed. The study of secondary consumers showed that carnivorous ants were most important, followed by spiders. Secondary consumers, however, required about twice the energy available in the productivity of the primary herbivores. This discrepancy may be partly explained by the problem of placing species in trophic levels and partly by inaccuracies of measurement. But these are insignificant amounts considering the amount of energy that was not assimilated or was deposited on the soil as waste, an amount of 70×10^6 kcal/ha/year. Since the savanna is, according to Lamotte, in an equilibrium and does not increase in biomass yearly, this material must be consumed in another way. Lamotte emphasized the potential role of microorganisms, fungi, bacteria, and actinomycetes in the savanna system, and estimated their potential metabolism at about 58 units. He concluded that the activity of these microorganisms is probably greater than that of all the animals in the system.

The Lamto study is interesting for several reasons. First, it represents a biological rather than a physical approach to an ecosystem study that was not driven by modeling. Thus, the focus shifted to animals, which are the most diverse biotic elements in the system and show the most differences among taxa. The French biologists were well aware of the currents in population and community ecology and cast their studies in the jargon of this research area. Yet they did not find a way to represent this biologically focused approach within the ecosystem context. Rather, they fell back on the trophic level concept and used the energy flow between trophic levels to represent the activity of the overall system. In doing so, they discovered that the complexity associated with the fauna was dominated by vegetation, on one side, and microorganisms, on the other. What exactly is the role of animals in these systems?

I have struggled with this question in reviewing similar data on the ecosystem performance of old-field systems in South Carolina (Golley, 1974). My suggestion, reasoning analogically from information theory, was that animals serve as thermostats or control mechanisms in ecosystems. That is, they consume plant production and use most of what they consume in metabolism. Large animal populations could, theoretically, reduce the carbon in the system and thus control carbon buildup. Plants also respond to herbivory with increased productivity so that the consumption impact might be positive in certain situations. Animals also pass carbon through their systems as waste and speed up the activity of decomposing organisms, thus speeding up the cycles of producing the essential materials required by plants. In this way, animals can either speed up or slow down carbon dynamics.

The Lamto studies represent a transition between the mechanistic eco-system projects of the IBP biome program and the biologically based ecological studies of the late 1970s and 1980s. They exemplify a general type of ecosystem project that characterized the European effort in the IBP and fulfilled more exactly the plan of work that was outlined in the original meetings of the terrestrial IBP program (see chapter 5). In doing so, they maintained the direction of ecological studies that respected the biological adaptations of species, but they did not find a biologically acceptable way to represent this complexity in an ecosystem context. A question remains: If the physical, me-chanical approach ignored the biological character of the species in the system and the biologically oriented species approach led only to the discovery of biological complexity, how could ecologists find a compromise between these two equally limited approaches?

The United States also supported a few ecosystem studies that were out-side the biome program and were biologically focused like those at Lamto. These IBP studies were concerned with Hawaiian island ecosystems and with a comparison of Mediterranean and desert ecosystems in California, Arizona, Chile, and Argentina. The Argentine project illustrates an ecosystem project where the term *ecosystem* was used in a generic sense to indicate that the ecologists were addressing numerous components of ecological systems with-out any attempt at an overall system synthesis. The comparative project was called the Origin and Structure of Ecosystems project, and it began by asking, Does the biota in climatically similar regions having different geological and biogeographical histories exhibit a convergence of structure and function? The research was structured in a three-dimensional matrix that contrasted two climates—the Mediterranean and the desert—in two regions with different biogeographical histories—California and Arizona—with Chile and Argen-tina.[5] Research was focused in the third dimension of the matrix on physical and climatic factors, production and energetics, patterns of resource use, adap-tive strategies, phenology and behavior, morphology and biochemistry, and taxonomic composition. This grab bag list of foci represented a cross section of the interests of the participants and the topics of current interest in the ecologi-cal community. Unfortunately, the studies did not proceed equally across the matrix; most attention was paid to the California-Chile comparison.

The Mediterranean comparative studies showed that convergence of struc-ture was most apparent for vegetation and became less significant for herbivores and predators, and that was there was less convergence as one moved up the food chains.[6] Evergreen trees with sclerophyllous, or high specific weight leaves, predominated in both sites with a Mediterranean-type climate. As the climatic conditions become drier, there was a progression of vegetation types from evergreen forest in the more mesic situation through evergreen scrub to a drought-resistant deciduous vegetation in the driest situations. Similar trends in the distribution of the soil fauna also were observed. For example, under Mediterranean climatic conditions soil organisms were found at unusually deep soil depths. For animals in general, however, different evolutionary histories resulted in different structures and functions in comparatively similar climates and habitats. The answer to the convergence question turned out to be: it depends! But these studies were productive in another sense. After the IBP, the research group continued their collaboration and then expanded to consider convergence and comparisons over all Mediterranean climates, including those of Europe, North Africa, South Africa, and Australia. The program has now become a scientific society and meets at regular intervals in various countries.[7]

The Origin and Structure of Ecosystems IBP program emphasized the biological species in ecosystems and represented the growing interest in the role of Darwinian evolution in the organization of populations and population-population interactions. It went further in this direction than the Lamto study, which had tried to consider the whole ecosystem as a unit of study. A divergence in the use of the term *ecosystem* was developing in these different ecologies. In the programs that focused on biological species, *ecosystem* was becoming a term used to express a point of view or a perspective from which a research question started. The term was not being used in an operational sense, as it was in the biome programs and at Hubbard Brook.

RESEARCH APPLIED TO NATIONAL NEEDS

In 1970, NSF established another new program that was labeled "Research Applied to National Needs" (RANN). This program was to deal with the criticism that basic research is self-focused rather than society-focused, being motivated by such personal goals as awards, professional advancement, and academic power, as well as a desire to know and satisfy personal curiosity. In a society facing many technical problems, it was considered essential to build a more direct link between research and application. RANN was expected to bridge the gap between basic research and its application in solving social and technical problems.

RANN was divided into several divisions. One division relevant here focused on the environment and was directed by Philip Johnson. Johnson had known Bormann and Likens when he was with the U.S. Army Cold Regions Laboratory at Hanover, New Hampshire. He had moved to the University of Georgia and began a Hubbard Brook–like ecosystem project at Coweeta Hydrological Laboratory. Johnson was recruited by NSF to head the Division of Environmental Systems and Resources of RANN. From 1970 to 1977 the division provided about $20 million for two hundred studies; however, most of the funds went to eighteen large studies.

These larger studies were grouped in a program focused on Regional Environmental Systems (RES) and were designed to address a growing problem in environmental management. The United States was faced with a contradictory state of affairs. First, public interest in the environment was at an all-time high. Formation of the Council of Environmental Quality, passage of the National Environmental Policy Act, and the formation of the Environmental Protection Agency all occurred at this time. Ecological and environmental research was active. The development of large IBP projects used the latest techniques in systems analysis and computer science. Nevertheless, few new concepts of analysis and management of the environment were being used at the local level, where there was a confusion of direction and purpose and even conflict over how to detect, correct, and manage environmental issues and problems. Further, the various governmental agencies that were expected to provide leadership were each going in a different direction and using different methodologies and approaches.

In this arena, RES proposed to fund projects that would use the advanced technology of systems analysis, incorporate ecosystem information and concepts, economic and social systems, and link up with users so that the technology was directly transferred into solutions to problems. These projects were to operate at large, regional spatial scales, and were expected to develop computer software for management and problem solving.

In developing projects, RES worked with many parts of the academic community. Johnson, as an ecologist, was familiar with the developments in ecological modeling and among the eighteen large projects were several headed by researchers from other ecological projects. These included Howard Odum at the University of Florida, Kenneth Watt of the University of California at Davis, Al Voelker of the Oak Ridge National Laboratory Regional Studies Group, Gerard Schreuder of the University of Washington College of Forestry, and Don Jameson of Colorado State University. Voelker, Schreuder, and Jameson were all extending IBP project experience into a new sphere. Thus, there was a direct connection between the RANN regional projects and the IBP biome projects.

After seven years of operation, the RES projects were evaluated by a team headed by Brian Mar of the University of Washington department of civil engineering.[8] Mar's review revealed a series of fundamental problems with the transfer of scientific information to problem solving. First, the concerns of the producer and user groups were very different. The two groups were widely separated in space, interest, and motivation. The academics were mainly interested in developing new techniques and concepts; they receive few rewards for applying these to problem solving. In contrast, the user groups were deeply committed to their specific problems and suspicious of generalists. They wanted the academics to work at the local level long enough to understand their problems and become part of a team. Overcoming these differences was difficult, and the projects had mixed success in bridging the gap.

In the early 1970s, the technology available to these scientists was limited. For example, a major issue was the decision to organize data spatially. If a spatial analysis was required, it increased the cost tremendously and made the project more difficult because geographic information systems were just being developed. Each project created its own modeling approaches and its own products, which were difficult to transfer and to evaluate. Personal computers were not yet in common use, and some discussion centered on the use of central versus minicomputers. There was difficulty in integrating data from the social areas, such as population and economic data, with the resource and residual areas that dealt with pollution and the ecosystem. Models in resource and residual management were the most applicable, since they had been developed by environmental engineers for specific purposes relevant to RANN objectives. Ecosystem models were still being developed and reflected the modelers' ecological theories. Population, transport, and economic models had different forms and were of different degrees of utility in the program.

Finally, there were problems with the form and accuracy of the data used in the models and the concepts that underlay the program. First, these investigators were faced with hierarchically organized systems in which each level was influenced by different environmental factors. For example, regional dynamics might be constrained by economic influences, while local issues might be controlled by social organizations. The data available at all levels of these hierarchies were not equally sound, and with regional systems one can not run experiments to test the validity of the data. Second, the theory used to organize data was usually deterministic, implying that the system should behave in a specific way under a specific set of inputs. Not only was this assumption under attack by ecologists and sociologists, but there was no way to determine whether the theory was correct. Because of this problem, local authorities were highly critical of academics, and conflict arose. We have learned that the application of scientific information to a problem is such an individual and specific process that it cannot be generalized successfully, except when dealing with physically constrained systems. As we move to biological and social systems, individual behavior becomes more important until it becomes the key to success or failure. The RANN program was developed at a time when ecologists were debating whether all systems followed physical rules and were basically deterministic. Now we recognize the relativity of systems and that they reflect the chance presence of individuals with specific abilities and motivations.

Mar's comment on ecosystem analysis summarizes the experience of RANN relevant to the development of the ecosystem concept:

> Mathematical ecology is a relatively new and developing field that received a major acceleration from the IBP efforts. The principles of energy and mass flows, ecosystem stability and resilience, and the linkage of compartments of an ecosystem are attractive concepts, but implementation and adaptation to urban systems, social systems, and regional systems must overcome the interdisciplinary coordination problems. There are relatively few ecosystem modelers with adequate training or experience to participate in such efforts. Of the modelers available, problems may ensue because they tend to have strong paradigms and do not yield well to conflicting paradigms of other disciplines.
>
> The philosophy provided by ecosystem analysis can be assimilated rather than substituted for more conventional RES methodology. Because experience is limited with these models, evaluation and selection is difficult to perform on a rational basis, thereby requiring more development and application to raise this modeling technology to the equivalence of physical models. Problem-specific research will be more effective in the short run, but models can be used in the research to probe system structure and dynamics. (Mar, 1977, 176–77)

These comments could have been applied to the IBP biome projects as well. His identification of the personal, idiosyncratic nature of ecological concepts and the difficulty of creating successful team work or integrative concepts among ecologists or across disciplines reflected an outsider's identification of a fundamental weakness of ecosystem studies at that time.

MAN AND THE BIOSPHERE

In 1971, UNESCO launched a new program that grew from the IBP experience. This program was called the Man and Biosphere program (MAB). It addressed several of the criticisms of the IBP. First, like RANN, it was practical and focused on solving specific environmental problems and transferring scientific knowledge to this problem solving at the international level. The lack of practicality had been a criticism of IBP by the ecologists of the USSR and the Central European countries, and while IBP was intended to understand the basis of biological productivity for human welfare, it was quickly captured by ecologists of dominant western countries who created an academic program that explored the reigning ecological paradigms. MAB reversed this emphasis, building upon the substantial advances in knowledge that came from IBP research. Second, MAB focused on developing countries and the ecosystems in those countries that had been neglected or underemphasized in the IBP. For example, MAB set research on tropical forests as its highest priority, with arid and semiarid grasslands and deserts as its second priority. Third, MAB studied systems in which humans were an integral part, including cities, agricultural systems, and natural reserves. In this emphasis MAB was following the same logical development of research as the U.S. RANN program, but without the heavy emphasis on system modeling.

The MAB program was endorsed by the United Nations Conference on the Human Environment in 1972 and began a period of defining its objectives and scope. The shift from the ICSU system, which had sheltered IBP, to the U.N. system was extremely significant because it meant a shift from academic science and individuals to governmental science. This shift permitted direct funding by governments, the involvement of governmental agencies in research, and access to multilateral aid programs, which could assist work in developing countries. UNESCO saw that committees designing and operating MAB had representation from all national perspectives. Although this is one of the costs of working within the U.N. system and reduces efficiency markedly, it also prevents too strong an influence by the economically or militarily powerful nations. When the problems concern the environment and social systems, this can be helpful.

A priority task of MAB was to select a director. The choice was a Venetian who had migrated to Chile, Francesco di Castri. Di Castri was a soil ecologist who had been one of the principal investigators of the Structure and Origin of Ecosystem program of the IBP, so he was familiar with the IBP. Di Castri was also an active, creative individual. He had revitalized Chilean ecology, organized several scientific associations in Chile, and established a wide range of international connections. He was able to express himself in several European languages and had a strong interest in applying ecology to solve problems, especially in the third world. Di Castri was an excellent choice, and the program was well led.

Di Castri put together a staff of scientists and communicators. Among these was Malcolm Hadley, an ecologist trained in Durham, England, who, after receiving his doctorate, joined Bourliere in Paris as assistant to the president of the ICSU/IBP steering committee. Thus, Hadley also had inside knowledge of the successes and failures of the IBP. The MAB staff worked inordinately hard to put together a program. Morale was high, enthusiasm was infectious, and the hours were long. It resembled the Hubbard Brook or Coweeta ecosystem projects, with enthusiastic, young, active scientists involved in important work. Di Castri protected his associates from the U.N. bureaucracy as much as possible and through personal diplomacy built support for MAB worldwide.

Such effort within a complex bureaucracy could survive only for a limited time, and after a ten-year review in 1981, the di Castri era began to end. Eventually, di Castri moved to head the CNRS center for ecological research at Montpellier, France, and Bernd Von Droste assumed the MAB directorship.[9]

The relevance of MAB for ecosystem studies was threefold. First, MAB provided a reason to extend ecosystem studies into areas that had been neglected or inadequately treated in the IBP. This was especially notable for tropical rain forests and arid regions. For example, Venezuelan ecologists Ernesto Medina and Rafael Herrera responded to a government request to understand the basis of productivity in the rain forest in southern Venezuela by establishing a MAB pilot project, called the San Carlos project. San Carlos is located near the Brazilian border on the Rio Negro River. The Venezuelan government was interested in establishing settlements on its southern border with Brazil, but various attempts at settlement had failed. The soils were exceptionally poor, yet they supported rich, tropical forests. The question was, how could these forest grow, yet settlers' crops, cattle pastures, and tree plantations did not grow on the same soils? Medina and Herrera invited ecologists from the Federal Republic of Germany and the United States to join them in an international ecosystem project to study nutrient cycling in the San Carlos rain forest. The project was especially successful. It led to many academic publications and answered the practical question behind it. The forests sustained themselves through the evolution of many nutrient conserving mechanisms, which resulted in almost closed cycles of nutrients.[10] The atmospheric inputs of essential chemical elements was sufficient to maintain nutrient stocks and forest productivity. But when the forest was cut, all of these mechanisms were lost, and humans could replace them only at great cost, with ingenuity and hard labor. The costs of replacing the mechanisms would far outweigh the income derived from the products of the system. The principal U.S. investigator, Carl Jordan (1987) of the University of Georgia, with his graduate students, extended these studies throughout the Amazon basin, creating an ecosystem-based approach to tropical rain forest management.

Second, MAB undertook to extend radically and improve on the conservation activity of the IBP. In this effort MAB established a Biosphere Reserve

Program, which identified natural or seminatural areas worldwide where research could be undertaken and the environment protected. The idea of biosphere reserves was to allow research and some economic activities in a portion of the area, while leaving other portions as undisturbed controls. This was an effective program that by 1977 had established over two hundred reserves in fifty-five countries (Maladague, 1984). Although there were widely different uses made of the biosphere reserves, the basis was laid for a worldwide network of sites for ecosystem studies that would lead to a comparative program of the ecology of the biosphere. This comparative aspect has never been realized, but the potential remains.

Third, MAB extended ecosystem studies from natural landscapes to the human-built environment, leading to a revitalization of the subject of human ecology on ecosystem principles. An example of a successful MAB project of this type was one developed in the Austrian mountain valley of Obergurgl (Moser and Peterson, 1982). Obergurgl is a small village with surrounding Alps at the end of a valley, which had become a popular mountain resort for climbing and hiking. In 1966, when I visited the area, there was still a mixture of tourist facilities and traditional mountain farming activity, with most of the tourist activity concentrated in the summer months. As recreational skiing became the dominant economic activity in European mountain regions, however, a variety of social, environmental, and economic conflicts developed. This was the same kind of situation that had motivated the organization of the RANN program; however, unlike the RANN program, with its emphasis on academic systems analysis, the Obergurgl program included local people in the design and analysis of the project. The consequence of this different approach was that the researchers had the social issues clearly defined for them; they had access to traditional knowledge of the region and a link to local decision makers. In this way, after appropriate ecological and social studies, a compromise was found between tourist activities and local, traditional needs, with preservation of the environment. Even though a solution of this type is seldom general and cannot be transferred to another area, the Obergurgl experience was sufficiently positive that the solutions and approaches were transferred elsewhere in the Austrian Alps to the general benefit of the region.

Thus, MAB led to a deepening and extension of the ecosystem concept— although it moved away from further theoretical evolution of the concept. Through MAB and the other activities discussed here, ecosystem studies gradually changed focus.

An examination of ecosystem studies that was not part of the active biome studies of the IBP shows that the ecosystem approach had practical utility in managing natural resources and in solving environmental problems. The need was great: there were no other obvious approaches, and ecosystem scientists were able to claim a practical use for their theory. Still, ecosystem modeling was not the route to successful application, just as it was not the route to a successful study of the biomes. Rather, successful applied ecosystem work followed the procedures of normal scientific work. An object or process was isolated and its pattern described by careful observation. The hypotheses erected to explain the pattern were then tested by further observation and experimentation. Each experiment eliminated uncertainty and gradually an understanding of the pattern became clearer. Each new system required the same process of observation, hypothesis, testing, and interpretation. Normal ecosystem sciences proceed piece by piece, step by step toward a deeper understanding of the mechanisms responsible for an observed pattern. The Hubbard Brook project showed the way to this approach and made clear that ecosystem studies pursued in this way were no different in kind from those in other areas of science.

If we think of ecosystem studies in this way, the development of theory and principles of ecosystem performance lies in the future as experimental science reaches conclusions about alternative hypotheses, implying that the application of ecosystem theory was premature in the late 1960s and early 1970s. Although the claim of relevance was used to justify funding the biome programs, there was a clear intention in RANN and MAB of applying the ecosystem approach. Success in solving environmental problems was achieved when specific local needs were made paramount and when a project was focused on specific goals.

The consequence of the massive application of staff, funds, and facilities worldwide on this scientific concept during the environmental decade was a sorting out of useful and effective methods and approaches to the study of ecosystems. The process led to a stage of maturity of the ecosystem concept and the development of normal ecosystem science. It led away from the theoretical, physical, energy theory of ecosystems built on an analytical system base that extended the machine metaphor into nature.

Maturity of the ecosystem concept did not solve the problem of linking ecosystem studies with evolutionary ecology, however. There was still no way to incorporate the knowledge of selection at the level of the individual or the specific behavior of populations and species into the ecosystem concept. There were too many organisms to consider. The constraints of the physical and biological environments on individuals were not adequately understood. Knowledge of the biology of organisms was too inadequate to manage the environment with confidence. Yet raising these questions within the context of ecosystem studies created further support and justification for biological studies of life history, natural history, and on the taxonomy of organisms.

In addition to the development of an operational approach to the study of ecosystems as objects in nature, the ecosystem concept continued to be employed as a point of view. When used in this way, the concept meant that the investigators had in mind that they were dealing with a natural system wherein the components were linked and interacted; this system had certain operating properties and controls and evidenced a constant pattern under certain conditions. It is a more modern way to speak of the wholes and connections of Elton, Clements, Tansley, and Lindeman because it is cast in the language of systems engineering and computers. This modern usage is metaphorical in that it recognizes that a process may be the result of many interactions and that an action can cause numerous interactions to reverberate throughout a system. Although it is not strictly a matter of common sense, as Eugene Odum (1977) asserted, the ecosystem approach does tend to include intuitive thinking, traditional wisdom and practice, and careful tinkering. Thus, in this latter sense, the ecosystem concept has become a general perspective or approach that signals that the individual using it has a broader vision, a holistic perspective, and is open to alternatives. In this sense, it finds extensive usage. As John Sheail (1987) commented, almost all recent presidents of the British Ecological Society have referred to the ecosystem perspective in their presidential addresses. Few have carried out ecosystem research.

In this formative period, when the ecosystem concept reached a point where it could become mature and be studied as normal science, the two elements that had been central parts of the concept became separated. The first element was the theory that ecosystems existed as discrete objects in nature, where they had a characteristic structure and function and could be studied directly. A scientific theory to explain and eventually predict their behavior was possible and would be of wide practical value. The second element was the point of view that the ecosystem concept was a useful way to think about the way the world is organized. It emphasized the interconnection and integration of systems at a variety of scales, cooperation, synergisms, and symbioses rather than dialectical opposition, competition, and conflict. If we adopt this latter point of view, we will manage our relations with others and with the environment in a different way than if we view humans and nature as separate systems. Thus, the ecosystem perspective can lead toward an ecological philosophy, and from philosophy it can lead to an environmental value system, environmental law, and a political agenda.

CHAPTER 7

Interpretations and
Conclusions

The development of a concept involves an interaction between ideas, the individuals involved, and the context in which the event takes place. In this final chapter I examine each of these influences to clarify our understanding of the ecosystem concept and how it developed. I place special weight on the contextual aspect of the analysis, especially the social context, because it has been neglected and is key to understanding ecological history. I then consider various challenges to the concept and the relation between the ecosystem concept, ecology, and science in general.

This story has been about the development of a concept that has provided order in the complex and multidimensional science of ecology. The search for order is a fundamental human trait and is found in all disciplines and cultures. Our concepts, technologies, and constructions may be interpreted as attempts to impose a human-created order on a complex and unpredictable environment that directly affects human well-being and survival. In this sense, a scientific concept has parallels to a road network. At a time of rapid change in human social relations or in human nature, we would expect a searching for forms of order that might reestablish social and environmental equilibrium. These searches are carried out within the appropriate cultural traditions of society.

The nineteenth and the twentieth centuries have been times of rapid social and environmental change. Nationalism, industrialization, urbanization, world wars, and economic depression characterize the period. It is not surprising that ecologists have been influenced by these trends. Nor is it surprising that ecological interpretations of natural order or disorder have become the foundation for concepts of social and psychological unity and order.

Before systems theory, cybernetics, computers, and their appropriate mathematics, ecologists described forms of order verbally and used simple diagrams to show interactions. Each scholar was inclined to coin a new word to represent his or her particular idea. Because a new, personal concept presented in familiar words might confuse listeners, scientists often used jargon to present their ideas. Haeckel's *ecology* was such a word, and the prefix *eco-* is now commonly used to form words related to environmental matters.[1]

Ecologists describing orderly aggregations of plants and animals in nature gave these unities special names, such as *ecosystem, biogeocenosis, microcosm, epimorph, elementary landscape, microlandscape, biosystem, holocoen, biochora, ecotope, geocenosis, facies, epifacies, diatope, and bioecos.*[2] Among these expressions, *ecosystem* proved to be one of the most useful for expressing the concept of a unity of interacting organisms and environment.

We can only speculate why. It incorporates the prefix *eco-*, which has been popular during this time of environmental concern; it emphasizes the system, which is both a modern and a technical concept; and it conveys the idea of an ecological machine. In western societies the machine metaphor has had strong impact and appeal.[3] The word *ecosystem* also is concise. Whatever the cause, Tansley's invention has had an appeal and success.

The entire pre–second world war period in which ecology became self-conscious and organized was a formative period for the development of the ecosystem concept. The precursors of the concept are found throughout ecology. Further, ecosystem-like studies were being carried out from the last decades of the nineteenth century. They simply were not given the name or understood within an ecosystem concept. Tansley's term, then, does not signal a new research area or a new form of ecological knowledge. Rather, it brought together extant knowledge, expressing it in a new way.

In the preceding chapters, I have described the development of the ecosystem concept in the continuous flow of time. I have suggested that the ecosystem concept developed in phases: a formative phase from the late nineteenth century to the second world war, an initial organizational phase from the second world war until the mid-1960s, and a rapid growth phase during the IBP. This chronological scheme provides a skeleton upon which to build a body of explanation. We can begin to do this by considering the context, or the environment, that formed this scientific concept. For example, we can ask why this concept arose in certain nations and not in others. What social and cultural factors shaped the concept? How did the development of ecology influence this particular ecological idea? We might also examine the ecosystem concept itself and determine whether it was complete and avoided internal contradictions. My purpose here is to build a broad explanation of why the ecosystem concept developed and matured in its particular form.

The Environment of the Ecosystem Concept

I begin by considering the spatial pattern of the development of the ecosystem concept. The concept developed in the northern hemisphere, in the USSR, Germany, the United Kingdom, and the United States. The northern hemisphere was also the heartland for development of ecology as a whole (McIntosh, 1985). Anna Bramwell (1989), a British historian, has implied that ecology has a relation with northern and Protestant countries, but this suggestion does not hold up under examination. Russia is hardly Protestant and the USSR and the United States include such a range of environments that the concept of northernness does not adequately describe their ecological place in the world.

Rather, it seems that each of these countries had specific reasons for contributing to the development of the ecosystem concept and to ecology in general. These reasons derived from the social, cultural, and historical factors operating in each nation. Of course, in a broad sense, the development of interest in ecosystems required an interest in nature and an open intellectual environment to allow new ideas about nature to be expressed. Yet this sort of generalization is not very helpful. Why, for example, did neither France nor Sweden, both large European countries that had made important contributions to ecology, play major roles in the development of the ecosystem concept? To find an answer, I consider the social-cultural context in which science evolved in each country and whether there were general patterns that distinguished those countries that contributed to the formation of the concept from those that could have but did not. In avoiding environmental determinism, I nevertheless want to consider the environmental-cultural context in concept development.

During the formative period for ecology and the ecosystem concept in czarist Russia, there was a reaction to industrialization, modernization, and political unrest that extolled traditions of Russian peasantry, the close relation of the peasants to the land, and the natural landscapes of the Russian forest and steppe. The most important figure associated with this form of thought was probably Leo Nikoleyevitch Tolstoy (Lavrin, 1946), even though the attitudes of this giant of Russian culture were echoed by many others. Tolstoy's mystical attachment to the peasantry and to the soil reflected a practical reality. Douglas Weiner (1988, 12) refers to this background in his discussion of the development of Russian conservation: "By the early 1890s, the rich practical traditions in agronomy, forestry, and meadow management in Russia had come together into a self-conscious science of phytosociology—the study of vegetation communities. Perhaps conditioned by the traditional Russian value of community feeling (*sobornost*), the pioneers of plant ecology looked to variegated 'virgin' nature as a model of harmony, efficiency, and productivity that the agriculturist should strive to emulate. To put agriculture on a truly sound basis, the early ecologists stressed, it was first necessary to study pristine natural communities—their origins, development, and spatial and temporal transitions." This form of thought was reflected by some Russian botanists and zoologists in their reactions to Darwin's *Origin* of Species, which had substantial impact in Russia. For example, the "father of Russian botany," Andrei Nikolaevich Beketov (1825–1902), wrote an essay in 1860 on the harmony of nature (Todes, 1989). In that essay, he emphasized the mutually dependent relations among organisms and physical conditions, which he attributed to the adaptation of every phenomenon to its specific purpose and to its environment. The zoologist Karl Fedorovich Kessler (1815–81) emphasized mutual aid as a contrasting principle to Darwin's theory of competition. Mutual aid became a key feature of Russian biological thought and is especially known through the writings of Prince Petr Alekseevich Kroptokin (1842–1921).

In contrast to this line of thought, soil science, geography, and geochemistry were being organized on modern lines. Vasilii Dokuchaev was the founder of scientific soil science. Vladimir Vernadsky played a similar role in geochemistry, being especially interested in the geochemistry of the biosphere. His student, A. E. Fersman (1883–1945), developed geochemical mapping. It was this combination of soil science, geochemistry, and geography that produced landscape geochemistry as a distinct field of Soviet science.

Thus, before the Revolution of 1917, there was in Russia both an intellectual population culturally receptive to integrated concepts of man and land, and scientific leaders with a broad understanding of natural systems. The Russian revolution unleashed a vigorous fountain of creativity—known from the perspective of the arts as the Russian avant garde. Release from the constraints of the bureaucracy and czarist culture produced a dazzling development in modern art, drama, and literature. The sciences were equally stimulated. Even Vernadsky, who was not sympathetic to the Bolsheviks, was enticed back from France in 1926. Finally, further support came from Marxist-Leninist philosophy itself, which, as Richard Levins and Richard Lewontin point out, stresses "the unity of structure and process, the wholeness of things, both between the organisms and their surroundings and within organisms" and "the integration of phenomena at different levels of organization (Levins and Lewontin, 1985, conclusion).

In the time from the end of the civil war that followed the revolution to the Stalinist period, there was an opportunity for the emergence of a modern, scientific, integrated concept of nature, which could have wide application in natural resource management, and such a concept did appear. Further, it anticipated Lindeman and Hutchinson by about twenty years in its focus on trophic dynamics and energy flow. We are in debt to Weiner (1988), who uncovered the story.

Among the professors trained before the revolution and willing and able to continue research and teaching under the new government was a remarkable individual named Vladimir Vladimirovich Stanchinskii. Stanchinskii was born in 1882 and obtained his doctorate in natural sciences from Heidelberg University in 1906. Before the revolution he taught in the zoology department at the Moscow Agricultural Academy. After the civil war, he was professor and head of the zoology department at Smolensk University. There his research focused on the mechanism of speciation, in which he hoped to unite the data of genetics with those of ecology and systematics. In 1927, he believed that he had answered the problem of speciation by positing a dual chromosomal material; one kind behaved according to Mendelian genetics and the other responded to environmental stimuli.

With the solution to this problem, Stanchinskii then turned to another central problem of biology, the nature of the biological community. In Stanchinskii's concept the organisms that made up the community were dynamic, continually changing, cycling materials and energy and were linked through an exchange of energy and materials. Each species had a specific biochemical and physiochemical role to play in the community and had a specific chemical structure. Even though these ideas about the exchange of matter and energy between living and nonliving nature were inspired by Vernadsky, Stanchinskii aimed to reduce biological phenomena to a common physical denominator, energy, and to express their relations in a set of mathematical equations, describing a relative stability or a dynamic equilibrium of the biological community. By 1931 Stanchinskii had worked out a simple mathematical model of the energy flow of a community, containing primary producers and a generalized heterotroph (Weiner, 1988, 80).

Of equal importance, Stanchinskii applied his community concept in the field. In 1929, he moved to the University of Kakhovka, near the important conservation reserve and agricultural experiment center, Askania-nova. Askania-nova nature reserve had been formed through the influence of the Polish ecologist Iosif Konradovich Pachoskii on a wealthy landowner, Fredrikh Eduardovich Fal'ts-Fein, who set aside five hundred hectares of virgin steppe on his estate (Weiner, 1988, 16, 71). Askania-nova was located at the mouth of the Dnepr River, near the city of Kakhovka, about a thousand kilometers south of Moscow. Under Stanchinskii's leadership, a new kind of research began. Censuses of insects and other animals were made in conjunction with determinations of vegetation biomass. The primary and secondary production of the species was determined, and the organisms were arranged in a theoretical trophic ladder, under the expectation, from the second law of thermodynamics, that energy would become progressively less available as one proceeded up this ladder.

In 1929, Stanchinskii was at the height of his career. He was not only head of the vertebrate zoology faculty at Kharkov State University, but he was also founder of the University's Zoological-Biological Scientific Institute and headed the institute's ecology division. He served as editor of the USSR's first ecology journal. He was asked to organize the fifth Congress of Zoologists in 1933 in Kharkov. Unfortunately for Soviet ecology, he was not fated to continue this leadership.

While Stanchinskii was forming a new approach to biological communities, ecological science and conservation were coming under attack from a group representing both the generation raised and educated since the revolution and utilitarians looking for deviations from Marxist-Leninist philosophy. These people were opposed to any science that implied that nature might have properties, such as equilibrium conditions, that could limit human manipulation of nature; to analogies between human societies and natural communities; to the use of mathematics in biology; and to the study of theoretical questions for their own sake. One of their first targets was the ecologists. The attack was initiated by Isai Izrailovich Prezent, an associate of the infamous Trofim Denosovich Lysenko, who later led the destruction of genetics in the Soviet Union. Prezent was a lawyer who graduated from the faculty of social science of Leningrad University in 1926 and then became a leader in the Leningrad Communist Academy. In 1931, he founded and led the *kafedra* or subdepartment of the Dialectics of Nature and Evolutionary Science at Leningrad State University.

In 1930, at the fourth All-Union Congress of Zoology, Prezent attacked Stanchinskii's theoretical talk on trophic dynamics. He even raised doubts that ecology was a science.[4] At the same time, Prezent turned against the geobotanists and the conservationists. Ironically, at the very time when Soviet ecology was making great strides—passing from the formative stage of development to the active growth stage of an ecosystem-like theory of natural systems that would have wide application—it was attacked by the Stalinists and destroyed.

Stanchinskii tried to reframe his trophic-dynamic aspect of ecology by deemphasizing its mathematical formulation and the equilibrium hypothesis and arguing that "the biocenosis, unlike a living organism, did not develop along a specific path as a result of preexisting genetic instructions. Rather . . . the biocenosis [was] a system of species whose mutual, historical adaptation to each other and to the abiotic environment they shared (and created) was dialectical, unplanned, unpredictable, and unduplicatable. The development of the biocenosis, like a kaleidoscope, was the product of the emergence of new evolutionary facts through this unending flux of interactions" (Weiner, 1988, 214–15).

Stanchinskii's turnabout was insufficient to save him. He was purged in 1934. A book summarizing his trophic-dynamic studies of steppe communities, which had already been typeset, was destroyed, and he was removed from all his academic posts. Trophic-dynamic work essentially disappeared from the research agenda of Soviet institutions, although Borutsky (1939) carried out studies of the profundal of Lake Beloie, and Ivlev (1945) studied the trophic dynamics of the Caspian Sea littoral zone. Possibly, their association with limnology, which had little connection with agronomy and terrestrial biological subjects, protected them.

Although the consequence of the Prezent-Lysenko attack on ecology was the termination of trophic-dynamic research, other investigations relevant to ecosystem studies continued. I have noted advances in research in the cycling of chemicals that continued within geography and geochemistry, which led to the development of phytochemical prospecting techniques for minerals. In the 1940s, the forest ecologist V. N. Sukachev, who had been active since the early decades of the century, advanced an ecosystem theory similar to that of Tansley. Sukachev, following Vernadsky, proposed the term *biogeocenosis* for the interacting complex of organisms and environment. He rejected the word *ecosystem* and other available terms in favor of a manufactured word that would emphasize the interaction of the biological and geological complexes to form a single entity. In a book by Sukachev and N. Dylis (1964, 26), biogeocenosis is defined as "a combination on a specific area of the earth's surface of homogeneous natural phenomena (atmosphere, mineral strata, vegetable, animal, and microbic life, soil and water conditions), possessing its own specific type of interaction of these components and a definite type of interchange of their matter and energy among themselves and with other natural phenomena, and representing an internally-contradictory dialectical unity, being in constant movement and development." The dynamic character of the biogeocenosis is emphasized by Sukachev (1960): "The biogeocenosis as a whole develops through the interaction of all its variable components and in accordance with special laws. The very process of interaction among components constantly disrupts the established relationships, thereby affecting the evolution of the biogeocenosis as a whole.

Therefore, investigation of the specific laws governing the evolution of each component does not preclude the need of studying the laws of evolution of the biogeocenosis as a whole. The latter is more than the single sum of its components and laws of development. The biogeocenosis as a whole has its own distinctive qualities" (Sukachev, 1960, 583). As I have commented (Golley, 1984a), this expression fits the Soviet emphasis on soil and landscape science and the philosophy of dialectical materialism. It is also a concept that uses classification as the main synthetic theme, even though Sukachev and Dylis do introduce cybernetics into the discussion.

Clearly, the attack by political opportunists and fundamental Marxists on ecological science twenty years earlier had had an effect. The brilliant possibilities released by their revolution were truncated throughout Soviet society, and even scientists concerned with broad issues tended small, isolated gardens. Although the problems of translating from one language to another may have retarded the flow of ideas from East to West, nevertheless, the concept of the biogeocenosis had little impact on ecological science outside the USSR. During the IBP, when the ecosystem concept approached a stage of maturity in the West, some Soviet ecologists applied the ecosystem concept in their work. This was especially true in Estonia, where IBP investigations focused on the spruce forest ecosystem and made substantial contributions (Frey, 1977, 1979, 1981). The formative ideas of Dokuchaev, Vernadsky, and Stanchinskii never played their appropriate roles in the development of modern ecosystem theory and practice. A tragic history and the chance occurrence of individuals with a distinct animus overcame a hospitable environment of culture and landscape and creative scientific leaders to frustrate the evolution of a scientific concept.

A second group important in the development of the ecosystem concept was composed of the German ecologists, including German-speaking individuals in the neighboring countries of Austria, Switzerland, Holland, Sweden, and Denmark. The German foundation for ecosystem studies was firmly rooted both philosophically and technically. The philosophical roots were derived partly from romantic, idealistic, and progressive currents in Germany, which were a response to the unification of Germany in the nineteenth century and its rapid industrialization and urbanization. As in Russia and elsewhere, there was a tendency to react to those events by focusing on holistic, rural, and traditional values.

After the first world war and the economic collapse of Germany that followed, however, forward looking viewpoints were replaced by a negativity that was a reaction to contemporary events. Of course, there was no single pattern in a chaotic situation, but one can find holistic and natural sympathies in the *wandervögel* (wanderbirds) youth movement, the various national peasant parties, the national socialist party, and other movements. Walther Darré, who in 1933 became Adolf Hitler's minister of food and agriculture, was an especially strong spokesperson for these ideas. Darré stressed that the strength of Germany lay in the relation between its Germanic peasants and the soil (Bramwell, 1985), a relation expressed in the political slogan *Blut und Boden* (Blood and Soil), which became a pillar in the justification of Nazi policy.

Obviously, the social and political attitudes that eventually became tenets of the national socialist movement (Gasman, 1971) provided support for ecologists and limnologists taking a holistic approach to natural systems. It also explains the vehement opposition of some German biologists to holistic thinking. Thus, Thienemann and his associates at Plön were in a comfortable situation. Although their system studies were criticized by some biologists, the political situation was increasingly supportive.[5] They made substantial contributions to a systems approach to limnology, which was extended to other fields, including forestry (Lemmel, 1939).

In the aftermath of the second world war, however, with Germany economically prostrate and the policies of the national socialists totally discredited, there were few ecologists interested or capable of carrying on the holistic traditions. Thienemann retired and was replaced by Harald Sioli, who had spent the war years in Brazil where he became an expert on the Amazon River, and then by Jurgen Overbeck. Overbeck and Sioli rebuilt and continued the limnology program at Plön with a new emphasis that included tropical limnology and ecology. In this emphasis, the German studies were almost unique, and they established the research base for our knowledge of the Amazon River system just at a time when international interest was turning toward Amazonian deforestation (Sioli, 1957, 1963, 1968).

Within German biology there was much outright hostility to holistic, ecosystem-oriented work. The research focus shifted toward ecophysiology, which is at the other end of the scale of ecological organization. It was not until the IBP that Ellenberg built a team to study the dynamics of forest stands at Solling, Germany (Ellenberg, 1971). These studies were successful, but they did not survive IBP. Rather, the research evolved into addressing applied issues such as the forest dieback.

In contrast, the German orientation toward the relations of land and organisms expressed as an integrated system found continued support within a new initiative in geography that also included ecology. The initiative was labeled *landscape ecology* by the German geographer Carl Troll in 1939 (Troll, 1971) and was based on his studies of aerial photography. Landscape ecology provided a technical basis for land planning and has been a useful tool throughout Germany as well as in the Netherlands and Czechoslovakia. Landscape ecology brings soil scientists, hydrologists, geomorphologists, geologists, and ecologists together to create a synthesis at a relatively large spatial scale. Landscape ecology is broader than the biological ecology of the past, and it has a specific applied purpose and function.

Thus, the German interest in ecosystem studies, while well founded scientifically and supported philosophically within German culture, nevertheless, did not survive the second world war. Indeed, past political connections and the idealistic implications of the ecosystem concept—especially as applied by the Green movement—have created strong hostility toward ecosystem studies by some German ecologists. Even though Germany has the capacity to organize teams of researchers and could mount ecosystem projects, because of the ambiguous feelings of many scientists they do not support a vigorous research program in this area. Only where an individual with an interest in ecosystems obtains a position of leadership, such as professor and head of a department, or where ecologists are studying serious environmental problems, has an ecosystem-oriented effort been developed.

The situation in the United Kingdom was fundamentally different from that in Germany or in the USSR. British ecologists had been at the forefront during the foundation stage of development of the ecosystem concept. Following the second world war, British ecologists were stimulated by the trophic-dynamics research of Lindeman and Hutchinson to explore this new way of approaching nature. Yet their enthusiasm gradually declined until, in the end, the ecologists of the United Kingdom were largely focused on other ecological questions. What were the factors creating this particular pattern of involvement?

The causes for British activity and inactivity in ecosystem studies are complex and multifactorial. At first, there was a positive response to the opportunities for applying trophic-dynamic and productivity approaches in terrestrial and aquatic ecosystems during the postwar initial growth phase of concept development. For example, Ovington (1962) proposed that the ecosystem was the unifying theme for the study of forests. Ovington, who focused on productivity and biogeochemical relations in his published work, visited the University of Minnesota, the location of Lindeman's study, and worked with Donald Lawrence on production ecology in 1957 and 1958. Director James Cragg was told that the purpose of the Nature Conservancy's Merlewood research station was the study of the ecology of the woods ecosystem (Sheail, 1987, 256–57). Charles Elton (1966) organized an "ecological survey" of Wytham Woods near Oxford University in which he applied his habitat and food chain concepts. John Phillipson (1966) was exploring ecological energetics at Durham. At the same time, Amyan Macfadyen (1962) used the system perspective in his studies of the soil biota. Kenneth H. Mann (1969) applied the ecosystem concept in his investigations of the Thames River, and in an oft-quoted article, D. F. Westlake (1963) compared plant productivity in various environments. British ecologists were at the center of development of the ecosystem concept as they had been during its formative phase.

Meanwhile, other British biologists and ecologists focused on population regulation and the interaction of populations to form communities. For example, David Lack (1965) emphasized the role of competition in the regulation of bird communities, and V. C. Wynne-Edwards (1962) presented a theory proposing that behavior was an important factor in population regulation. A little later, just after this story of the ecosystem ends, John Harper (1977) extended these ideas to plants and created plant population ecology. The study of species populations and the interaction between populations captured the attention of ecologists and became a major area of British ecological work. It was not only exciting in itself, but it satisfied the scientific desire for ecology to move to a phase where hypotheses were tested through experiment and observation. One could undertake experiments with populations, and it was possible to apply the hypothetical-deductive approach to them. Further, species population ecology built upon the long British history of fieldwork in natural history in which botanists or zoologists collected, described, named, and reported on the distribution and abundance of organisms (Berry, 1988). Large research teams were not required. The research could be carried out by an individual ecologist.

Certainly, renewed interest in population ecology was one reason why British ecologists did not continue to play a major role in ecosystem studies, but this factor does not fully explain the move away from ecosystem research. Ecologists from the United Kingdom were active in organizing the IBP, and the plan for the terrestrial productivity program fit the research approach of the British ecologists mentioned above. They also were active in the international tundra studies, and they mounted, with Malaysia and Japan, a remarkable study of the Pasoh rain forest in Malaysia. In 1973, the British Ecological Society, with the British national IBP committee, sponsored a symposium to mark the formal end of the IBP. John Sheail (1987, 220) commented that it was one of the most ambitious, expensive, and concentrated programs of research ever attempted in Britain, and it produced a new type of ecologist who had learned to work as part of a team. During this period of ambitious work, however, many of the leaders of British ecosystem studies emigrated from Britain. For example, Mann and Cragg moved to Canada, and Ovington moved to Australia. Several leaders also retired. These included Elton, Lack, and George Varley of the Oxford University entomology department. The younger generation of ecologists tended to move away from ecosystem studies and toward population ecology. Nevertheless, in the British Ecological Society's survey of ecological concepts, the ecosystem concept was ranked first in importance by British ecologists in 1987 (Cherrett, 1989).

Finally, we turn to the United States and consider why ecosystem science developed so vigorously there. Again, we find that the cause is multidimensional. The United States responded to the economic depression of the 1930s and the second world war by forming centralized government agencies that could provide services that could not be, or would not be, supported through private or local initiatives. Following the war, Vannevar Bush, who had headed one such central agency, the Office of Scientific Research and Development, wrote a report on postwar science, *Science: The Endless Frontier,* at the request of President Franklin Roosevelt. Bush proposed the formation and funding of a national research foundation. This was the first step toward the organization of the National Science Foundation in 1950. Ecological research, which at first had been funded through the Office of Naval Research, now was supported also by the NSF biology program. In addition, other agencies with a concern about the environment began to support studies on ecology. The most important of these, as we have seen, was the Atomic Energy Commission. These research funds were immensely stimulating, since they provided support for graduate students as well as principal investigators, supplies, travel, and equipment. Funding was given to individual scientists through the institutions where they worked, therefore a competitive relation developed between scientists on a national scale. Local scientific work was deemphasized, except in applied fields, because most states could not or would not supply the funds necessary to support basic scientific research.

Because of these decisions and the federal funds that became available, there was a substantial increase in the number of researchers and scientific reports. Universities and colleges also recognized that research was an important complement to their missions. Many new academic and nonacademic institutions were founded that offered positions to graduates in the sciences. There was a relaxation in the hierarchical order of U.S. academic society, which was always much less rigid than that in Europe. This allowed individuals and institutions to obtain support for research and thereby challenge the dominance of institutions where ecology had been most actively studied before the war. In addition, these funds permitted the development of "big science" projects. Most of the projects were in physics and space science, although other fields shared in the opportunities. The IBP was big ecological science (Blair, 1977). At that time, it was not possible for any other country to organize or fund the great number of teams of ecologists researching basic questions. It was a special situation, and it permitted ecologists of the United States to dominate the growth phase of this form of ecological science.

In addition to these structural and financial advantages, there were organizational patterns within academic institutions that enhanced the opportunity for ecosystem research. Ecology could be taught and studied in many areas of the universities, but ecologists were primarily located in biology, botany, or zoology departments. After the war, biology departments continued to be organized around subjects such as anatomy, taxonomy, genetics, and physiology. Ecology was taught as a field course, with an emphasis on taxonomy. The excitement for students of biology was in the unfolding discoveries of DNA and the genetic code and the promise in the development of cellular biology and biochemistry. In this context, ecology seemed focused on old-fashioned questions.

This teaching situation was profoundly changed with the publication and widespread adoption of Odum's *Fundamentals of Ecology* in the early 1950s. Students then had the ecosystem concept as the organizing theme for ecology. This theme was new and dynamic and linked ecology to advances in physics and chemistry. Even so, it was difficult for many teachers to adapt to this new way of approaching ecology. Summer teaching institutes, funded by the National Science Foundation, were organized to train college instructors in the new ideas. Ecosystem studies were able to grow within biology departments because many of the models and methods of animal and plant physiology could be carried over into ecology. This was especially true of studies on photosynthesis and metabolism, both essential for studies of energy flow. Even as early as the mid-1950s ecologists were using primitive carbon dioxide absorption towers in the field to measure the photosynthesis of vegetation and were placing organisms in respirometers and metabolic chambers. These studies matured into a distinct subdiscipline of ecology, physiological ecology. The machinery and techniques of physiological ecologists have become immensely sophisticated, but in the United States, at least, they still retain their close connections to ecosystem studies (Mooney, 1975).

The consequence of organizing ecosystem studies within biology, however, was a deemphasis on the physical-chemical environment, which was one-half of the paradigmatic system. Biology students were not encouraged to take physical chemistry, geochemistry, hydrology, geomorphology, geology, or atmospheric physics. Indeed, ecologists were still required to have a command of the biological premedical or botanical curriculum. Thus, ecosystem studies developed in an unbalanced fashion with too strong an emphasis on the biological aspects and too little understanding of the environment. This bias contributed to the decisions made during the IBP to emphasize the modeling of biological components.

Why then did ecosystem studies not develop in the geographical, earth, or atmospheric sciences? Certainly, there were individuals within these disciplines aware of and interested in ecosystem studies, but the overly strong biological focus and the abstract nature of ecosystem studies in space and time provided few questions that required physical-chemical knowledge. This point was noted and criticized by physicist David Gates and others,[6] but the earth and atmospheric sciences were focused on their own questions, and there was no point of convergence. Further, geography was badly fractured into many subdisciplines and had little status in the academic hierarchy. As a discipline it received little direct support from NSF.[7] Thus, although it should have, it did not contribute greatly to the development of ecosystem science.

The applied fields with a focus on natural resources, such as agriculture and forestry, were initially hostile to ecosystem studies. These subjects seemed to be closely tied to client groups in the private sector, and they reacted to the growing environmental awareness in the United States stemming from Carson's *Silent Spring* by adopting a generally hostile position. Rather than adopt a strategy of responding to public concern and using this concern to generate new research opportunities, many of the practitioners of these disciplines dug in and considered themselves to be at war with the growing environmentally concerned citizenry. It was only after public laws mandating environmental study and action were passed that these disciplines began to adopt a broader interest in ecology. With the gradual retirement of the postwar generation of agricultural and forestry professors and scientists, this pattern began to change. Yet as environmental studies began to expand, it meant that the natural resource disciplines tended to be opposed to ecosystem research rather than supportive of it. In contrast, the ecologists, propelled by their broad ecosystem concept, were in the center of the environmental movement. The focus on environment, coupled with the student political movements of the late 1960s, created a social ferment that contributed to the phase of rapid concept development. The IBP was only one expression of this social-political activity. Thus, the nest for nurturing ecosystem studies was in the basic biology departments of universities and colleges, but the engine that propelled it was a social-cultural phenomenon, fueled by adequate financial and intellectual resources.

Other nations organized ecosystem studies, but most were unable to sustain them. France, Japan, and Sweden had traditions in biology, particularly in field biology, yet their academic science programs were highly structured and dependent upon the interests of those individuals filling professorial chairs in botany and zoology, subjects that might contain ecology. In these three countries there were few professors supportive of ecosystem-like studies during any of the formative or initial growth years of the concept.

During the IBP, France mounted an ambitious program of research that had projects both in France and in African countries, indicating that French ecologists were well aware of the opportunities for ecosystem studies and could organize such studies if financial and administrative support was available. Scientific leaders such as Bourliere, Lamotte, and others were present to organize and direct ecosystem projects. These French studies tended to be less exclusively focused on traditional topics of ecosystem studies, such as energy flow, biogeochemical cycling, and ecological modeling. Rather, they included population and community ecology and used productivity as an integrating concept. The flavor of this research came from the zoological orientation of several of the leaders. When IBP ended, the projects slowly declined, and ecosystem studies continued only at Montpellier, Toulouse, and a few other centers, where they gradually shifted toward being dynamic analyses of landscapes. In general, the ecosystem concept has not been a strong organizing principle in French ecological science.

As di Castri (1983) observed in his important review of ecology in France, the characteristics of French university organization and governmental research tend to maintain ongoing structures and programs and to be inflexible in supporting innovative or unconventional programs. This tendency makes it both difficult to organize new initiatives outside the regular pattern of administrative structures and to bring people from several organizations together to focus on problems that require an integrated approach. Di Castri uses rather strong language in his evaluation, ending with a discussion of the "blockage responsible for the crisis of ecology in France." Among these barriers, di Castri identifies problems in training students in fields that cross academic and practical subjects, in recruiting scientists and maintaining a scientific career, inadequate funding, and the lack of reasonable standards of performance applied consistently. Clearly, these comments suggest that there are various structural problems in French academic institutions and that these problems have limited the development of ecological science, including ecosystem studies. Almost ten years later, J. C. Lefeuvre (1990) repeated these criticisms.

Nevertheless, it is not entirely clear why France has not sustained the ecosystem studies organized under the IBP. In conversation, French ecologists cite the domination of the Braun-Blanquet approach in vegetation studies— which seems to parallel the impact of Claude Bernard and Louis Pasteur on French biology—as being responsible for the delay in French involvement in ecosystem work in its early stages. They also cite the strong division between basic and applied sciences and the difficulty of finding a common ground for sustained joint work. Finally, they cite the frustration, identified by di Castri, with the immobility of the academic structure. For ecosystem studies to prosper in France, it requires a professor committed to the subject, one who has the ability to create resources to support a research team over a long period, and who has the personality to attract to the research objectives a diverse collection of individuals from traditional subjects. Individual scientists have understood the challenge of this line of research and have made substantial and important contributions to it. Yet individual efforts cannot be converted into a national program without administrative commitment and support. The situation operative during the formative years of development of the ecosystem concept appear to be similar to those of the more recent period criticized by di Castri and Lefeuvre. France is not alone in having this problem.

Although Carl Linnaeus wrote about the economy and household of nature in eighteenth-century Sweden, his early contribution was not followed by a flowering of ecological science. As elsewhere, the development of Swedish ecology was limited by the organization of the universities (Soderquist, 1986). An emphasis on ecology depended upon the interests of the professors who occupied the chairs of zoology or botany, and naturally not all professors had an interest in ecology. It was therefore difficult to sustain the development of ecology within Sweden and required four generations of students before, in the 1960s, ecology was established as a continuing subject. Nevertheless, some Swedish ecologists, such as Einer Naumann, made contributions to ecosystem ecology in its formative period of development.

According to Thomas Soderquist (1986), the institutionalization of ecology was started with the publication of a Swedish edition of Odum's textbook by Carl-Cedric Coulianos in 1966. At that time in Sweden there was a convergence of the mass enthusiasm for nature stimulated by nature training in secondary schools, the results of scientific work in ecology by the fourth generation of ecologists trained in Sweden, and the widespread public concern over the deterioration of the national environment. This convergence resulted in governmental action on ecology, expressed in the context of the ecosystem. An ecological committee was formed in 1967 with Per Brink of the University of Lund as chair, and national policies of planning and resource management were placed on an ecological basis.

The IBP provided a stimulus for development of large-scale ecosystem projects in Sweden. Three projects were begun—one at Lund on deciduous forests, another at Uppsala, and a third focused on the Baltic Sea organized in the Asko laboratory of Stockholm University under Bengt-Owe Jansson.

In Soderquist's words (1986, 273): "The ecosystem projects stood out, in the minds of the leading science reformers, as the first serious attempt to organize a true scientific ecology. Ecosystem theory seemed to give to ecology a set of concepts and ideas, which was its own and not borrowed from other disciplines. This helped to establish the status of ecology as an independent scientific discipline rather than a speciality, whether of botany or zoology. The theory of the ecosystem thus accentuated the theoretical and scientific character of the new scientific order."

The Swedish ecosystem studies gradually became less active after the IBP, and the main emphasis in ecology was changed to population and community ecology. Soderquist attributes these changes to shifts in cultural attitudes in Sweden.

Japan poses yet another example. Plant ecology developed in Japan when Manabu Miyoshi returned in 1895 after spending three years in Germany with Wilhelm Pfeffer at the University of Leipzig.[8] Animal ecology was introduced to Japan by Tamiji Kawamura, who had studied with Victor Shelford at the University of Illinois, and at the Universities of Michigan and Cornell. In the post–second world war period Japanese education was changed along the lines of the American system, and many new laboratories were founded. Even so, ecosystem studies were not an important aspect of the work, even though the first edition of Odum's *Fundamentals* was translated into Japanese, and Odum toured Japan and lectured at Japanese universities in 1962.

The primary Japanese contributions to ecosystem studies came from their work on primary production. Masami Monsi and Toshiro Saeki, at the University of Tokyo, organized research on production ecology that led to a theoretical understanding of the comparative production ecology of plant communities. Tatsuo Kira and his colleagues also carried out studies on the production ecology of tropical rain forests in Cambodia and Thailand, which provided some of the first estimates of rain forest metabolism and standing crops. Thus, by the time of IBP there was in Japan a core of experience that allowed Japanese ecologists to organize a variety of ecosystem projects. Approximately six hundred scientists participated in the IBP, which resulted in twenty printed volumes reporting the research.

Most of the Japanese IBP projects were not ecosystem studies. Those that did take this perspective went deeply into an analysis of the system. For example, the report of the program that studied alpine coniferous forests at Shigayama, edited by Y. Kitazawa (1977), consisted of chapters on the structure of the biotic community, primary and secondary production, the decomposition of organic matter, and ecosystem metabolism. These divisions of the work follow the conventional trophic levels of the period.

Although the Japanese IBP ecosystem studies provided useful comparative data and extended the geographic spread of ecosystem studies to northern Asia, they did not continue after the end of the IBP. Makoto Numata (1990), quoting S. Mori's (1979) analysis of the state of Japanese ecology, commented that "the ecosystem approach is generally not very strong" in Japan.

The Social-Cultural Influence

It is clear why ecosystem studies developed and matured within the scientific community of the United States. They prospered, with other sciences, from the centralized funding and organization that had been created to deal with the problems of the great depression and the second world war. Ecosystem studies built upon a strong tradition in ecology, especially in the study of ecological succession and in limnology, and could extend the concepts of plant and animal physiology, which were dynamic elements within the biology curriculum, to field studies and could use the growing knowledge of systems science and computers. Finally, it was stimulated by, as well as contributed to, the widespread public concern about the environment, social justice, and the ending of the Vietnam War. Thus, ecosystem studies grew as part of a scientific advance that paralleled the military power and dominance of the United States internationally and a social-cultural countermovement that stressed environmental protection and social justice.

In contrast, the opportunities for the development of ecosystem studies could not be realized as quickly in other countries. The USSR had destroyed its capacity to contribute directly to the development of ecosystem studies, although important contributions were made in potentially important, but peripheral, areas such a geochemistry and landscape ecology. Germany was delayed in its capacity to contribute by the war and then, in the war's aftermath, German biologists were careful to distance themselves from holistic ideas that had some attachment to national socialism. Rather, an ecosystem orientation surfaced in landscape ecology, which was a joint enterprise of ecology and geography and provided the theoretical basis for landscape planning and design. In the United Kingdom many individuals made important contributions to ecosystem science, yet there was not the will, organization, or possibly, the support to maintain large-scale ecosystem projects. Other nations that might have contributed to development of the ecosystem concept were prevented from doing so by the organization of science in the universities and government. In most countries it was difficult to fit ecosystem studies into the structure of universities and research laboratories. The IBP allowed for the formation of large-scale ecosystem projects outside these institutions, which attracted widespread attention because of their size and newness and their potential to solve problems in a deteriorating human environment.

Yet the IBP had a finite life, and when it ended many of the ecosystem projects were bought to a close and published. The only country where this did not occur was the United States, where the funding was reprogrammed within NSF to support continuing ecosystem projects. The consequence of this decision was that ecosystem studies were institutionalized in the United States and became a recognized part of the scientific community. Nevertheless, ecosystem studies seldom became established within universities. This was less true at those institutions that had been centers of the IBP ecosystem projects. Later, in 1979, ecosystem studies were expanded further through a focus on long-term ecological research in NSF.

Certainly, this explanation, using national, historical, and cultural differences, is unsatisfactory because it is so complex and depends so strongly on the correlation of events. It may remind us of environmental and cultural determinism. Although language tends to emphasize cause and effect (one cannot qualify every statement and still tell a useful story), my conclusion from these comparisons is that social, economic, and political factors were of prime importance in creating an environment in which the ecosystem concept developed. These factors must be considered along with those involving the characteristics of the individual scientists who contribute to concept formation. In every country, the development of ecosystem research has depended upon individuals with imagination capable of providing leadership and direction. Yet without supportive social and political settings, they could not have developed ecosystem studies. Besides these factors, yet others may be relevant to understanding the development of a scientific concept. For example, the character of the concept itself: What was actually proposed? In its evolution, did the concept develop theoretical independence so that it structured a line of research? What are its implications and contradictions?

Ecosystem Theory

To consider these questions, I return to certain aspects of the theoretical development of the ecosystem concept. In tracing the dynamics of concept development to about 1975, I have concluded that the ecosystem concept went through three stages: a formative period, an initial growth stage, and a time of rapid growth. This conclusion makes sense from social-cultural and environmental perspectives, but does it hold when we focus on the theoretical aspect of the concept? We can begin by asking whether there was a theory of the ecosystem.

First, What do we mean by ecosystem theory? The word *theory* has several meanings: A theory may be a conception, a system of ideas and statements, or a hypothesis that has been tested and confirmed. The fundamental elements of a definition of theory focus on generality, on a foundational and systematic character, and on the process of confirmation. A notion, an analogy, or an untested hypothesis is not theory.

In the philosophy of science, there are several paths leading to the development of theory. For example, we may observe nature and find patterns that are consistent over space and time. The observation of consistency leads to a hypothesis that regularity exists. Further observation is used to test whether these patterns continue, especially under new conditions. If they do, we may form a theory about the regularity of the pattern. This form of reasoning comes from Francis Bacon. Bacon, in reacting against the medieval concept of deducing truth from divine authority, accepted experience as evidence of true knowledge. He called the procedure of passing from observation to hypothesis—to the test of a hypothesis by further observation—*induction* (Briggs and Peat, 1984, 18–21).

Karl Popper (1959), an Austrian-born philosopher educated in the 1920s and 1930s and working in Britain, proposed a radical alternative to induction. Popper noted that the scientist, as an objective observer, is not separate from the things that are being observed. Therefore the scientist can influence the observation. In order to retain objectivity, Popper developed what is called the *hypothetical-deductive* approach. According to Popper, theory is used to obtain answers from nature. Since theory spells out the observations and conclusions that can be tested experimentally, the theory stands or falls on the basis of such tests. Rejected theory is discarded. Thus, in the Popperian system, we proceed with a question or hypothesis derived from experience and test the hypothesis through experimentation. If it is false, the hypothesis is rejected, and we ask another question. A positive test, however, does not prove the hypothesis true. The classic example is that we observe a hundred white swans on a pond. Although we can hypothesize that all swans are white, a single black swan destroys our hypothesis. Science, in Popper's hypothetical-deductive approach, does not prove theory, it falsifies it.

Popper concluded that there were four elements of a good theory: (1) the elements do not contradict one another, (2) the conclusion is not buried in the premises, (3) the theory is simpler and more direct than alternative theories, and (4) it is capable of being falsified. These ideas of Karl Popper strongly influenced working scientists entering ecology in the 1970s (Peters, 1976; Saarinen, 1980).

Our conceptions about scientific theory were also altered by Thomas Kuhn (1962), a historian of science. Rather than theories, Kuhn identified paradigms in science. Paradigms are something like road maps, which structure how we see the landscape around us. The paradigm organizes what we observe and how we express our interpretation of the observations. Kuhn observed how scientists worked and found that Popper's falsification standard was a myth. Few scientists see anomalous experimental results as a challenge to the reigning paradigm. Rather, these kinds of results are interpreted as evidence that something was wrong with the methods, the instruments, or the way the question was being asked. The paradigm dominates the way scientists view their activity.

In the Kuhnian interpretation, theory must fit into the paradigm. Evidence that does not fit the paradigm is not accepted by most scientists, but if this evidence accumulates, some scientists become motivated to attack the paradigm. Usually, an attack is unsuccessful until the paradigm custodians die or retire, leaving the field to the critics who create a new paradigm. Kuhn calls this overturn of paradigms a *revolution*.

From Kuhn's perspective, it is clear that there is nothing like scientific progress, nor does science find truth or absolute knowledge. Rather, science addresses socially and culturally relevant questions within established paradigms. In other words, Kuhn believed that science is like other human activities, in that it serves a social good appropriate to time, place, and culture.

There is yet another view of scientific theory relevant to our understanding of the ecosystem concept. Physicists such as Werner Heisenberg and mathematicians such as Kurt Godel have shown that there are no fundamental axioms from which all others may be derived, nor is there absolute certainty about the nature of matter. Rather, at the most basic level, there is uncertainty about the nature of the world. Nature is probabilistic in that we can predict with accuracy the average pattern or condition, but we cannot predict the location of any individual element. This means that not only is the scientific observer conditioned socially and culturally, as Popper and Kuhn suggest, but the nature of matter is uncertain as well. This creates what John Briggs and F. David Peat (1984) have called "a looking glass world," from Lewis Carroll's book *Through the Looking Glass*. The fixed certainty of the past has dissolved into a reality where everything is in flux and everything is connected.

Thus, in this brief discussion, we have viewed a continuum from deterministic to relative theory. In ecology, theory has ranged across this continuum. Deterministic theory has been strongly associated with vegetation classification and ecological succession, especially the version developed by Clements. Although Henry Gleason also put forward a relativist theory of vegetation development, it was first rejected. The period from the second world war and the present has been a time of transition from an emphasis on determinism to an emphasis on relativity.

Given these alternatives, the ecosystem concept seems to fit best Kuhn's idea of a paradigm. It has been an overarching and organizing idea that was important in shaping ecology, especially in the United States, from about 1950 to 1965. The idea of an interacting system of organisms and environmental factors, organized into trophic levels or into food webs and food chains, linked through the flows of energy, dominated the science during the period. The influence of *Fundamentals of Ecology,* was important in moving the ecosystem from being one of several scientific concepts to a dominant position. It was not, however, a single, coherent, tightly reasoned theory, and I have purposely avoided the word *theory* to refer to it. Instead, I have used the less precise word *concept,* which implies a much more tentative and looser generalization. Rather, a broad concept such as *ecosystem* may hold a variety of ideas within it. The term *paradigm* seems to fit this broad generalization.

Even though we are using Kuhn's language, the development of the ecosystem paradigm does not, however, fit well into his scheme of scientific revolution. One reason for this lack of fit is that ecology itself was emerging as a science prior to the formulation of the ecosystem concept. In this long formative period of almost a hundred years, ecosystem-like concepts, together with other ecological concepts, were being developed. If there were dominant ideas in the period—in the United States at least—they included Clement's superorganism and monoclimax. A perusal of the pages of *Ecology,* however, shows that ecologists investigated all aspects of nature from many different perspectives during the period.

After the second world war ecology became more like a discipline, and its identification as a distinct subject was assisted by the ecosystem concept. Scientific ideas associated with the ecosystem concept could be reduced to physiology, chemistry, and physics. The ecosystem concept was also important in causing a convergence by transcending the dichotomies that had structured thinking, such as terrestrial and aquatic habitats, vertebrates and invertebrates, field natural history, and physiology. This integrating function, however, was also responsible for the incoherent and even contradictory nature of the paradigm. In their enthusiasm ecologists tended to load it too heavily and make it carry problems it was not designed to solve. I have used Levi-Strauss's term, *bricolage,* to describe the nature of ecosystem studies at this time.

At this stage the approach to research was shaped in part by the analogical thinking of several key individuals, including the Odum brothers and Ramon Margalef. These scientists tended to think in the form of analogies—such as, if the world is a heat engine, then . . . . This type of thinking was most useful in the initial growth phase of concept development; it became less useful as the concept matured, and eventually analogical thinking was strongly criticized.[9]

The period when ecological modeling dominated ecosystem studies, in the rapid growth phase, can be interpreted as a continuation of the analogical approach but in a more highly organized form. At first, ecologists (H. T. Odum, 1960; Margalef, 1962) used an analog of electrical energy networks to represent the energy flow pathways of ecosystems. By the 1960s, analog computers were being used in both teaching and research.

Analog computers were replaced by digital computers, as digital machines became more complex and could manipulate and store large amounts of data. The mathematical model, in the digital computer, became the ecosystem analog. Differential equations represented relations and flows, and the model structure was an analog of the ecosystem structure. Faith in these methods influenced the individuals who dominated IBP biome program planning.

It was not until the ecosystem modeling approach failed to provide adequate whole system descriptions and testable predictions that most ecosystem ecologists adopted reductionistic methods in ecosystem studies. The ecosystem model continued to be employed, especially by the generation that developed the concept, but it was no longer the dominant method used for structuring research questions. Instead, models became tools used for organizing data and generating questions.

Of course, a system's behavior can never be fully determined, because it entails many interactions, and there are always stochastic elements to be considered. The only way to improve the predictability of a system's behavior was to take the system out of nature and confine it in a controlled environment, where it could be manipulated experimentally. Ecologists used microcosms and mesocosms to study ecosystems in this way, which solved one problem but created another. The isolation and closure of the ecosystem ignores the role of other systems that control system behavior and introduce stochasticity. Ecosystems are open systems. Thus, neither the laboratory nor the modeling analogies led to the anticipated advances in knowledge of the ecosystems.

As the field of ecology matured, paradigms and theories appropriate to other levels of organization and to other question areas developed alongside the ecosystem concept. Although there is a tendency to think of ecology as a subject with a single paradigm, actually in ecology multiple paradigms coexist. This was Tom Fenchel's (1989) point in his answer to H. J. Carney's call for a integration of ecosystem and population ecology. Fenchel argued that there are many kinds of ecology. Indeed, he would like to give the word to the environmentalists and refer to each part by its name—biogeochemistry, population ecology, and so on. This is a specific example of a general pattern where the zones of interaction between the traditional disciplines have become the centers of new and interesting fields of study. It is for this reason that I have not contrasted ecosystem ecology with population or evolutionary ecology, as have others. Nor have I used a competition model and suggested that after a contest the ecosystem concept was replaced by some other ecological generalization. There has been no Kuhnian revolution, with the replacement of one generation by another. From my perspective, all the concepts coexist, and all contribute to the general knowledge of modern ecology.

Besides the theoretical and paradigmatic character of the ecosystem concept, there is another aspect that must be considered in the evaluation. This feature concerns the ecosystem as an object of scientific observation and study. In Tansley's conception, the ecosystem was clearly a physical object, located in a hierarchy of such objects. Yet, in much ecosystem work the ecosystem was a theoretical paradigm used to organize research and was not considered to be an object in nature. It is to this alternative perspective of the ecosystem that we should turn.

The technical term for the ecosystem as an object is *ecotope*. This term is derived from landscape ecology and refers to the smallest spatial object that has homogeneous properties. For example, at Hubbard Brook, Coweeta, and H. J. Andrews experimental forests, ecologists have demonstrated that small watersheds exhibit homogeneous water flow downslope and across weirs placed across exiting streams. These small watersheds can be defined accurately, experimentally manipulated, and studied like any other physical object. The watershed ecotopes are embedded in landscapes, and they individually contribute to the dynamic behavior of the higher order system.

Ecosystem research has matured and become widely useful through the study of forested watershed ecotopes and landscapes. These physical systems of relatively large spatial scale are relevant to environmental management, and therefore ecosystem research has become of value to land management agencies and other institutions concerned with the environment.

Finally, there is a third way in which the ecosystem concept has served science and society. In this usage, the concept serves as a bridge between a scientific paradigm, a physical object, and a holistic point of view. Over the past several hundred years, there has been a continual questioning of the modern synthesis that describes a deterministic order in the natural world and in human society. I have noted the evidence for this critique in science in the first decades of the twentieth century. The critique, labeled postmodern, has advanced a view of the world as dynamic, contingent, and changing. One direction for this perspective has been the view that systems are chaotic. A contrasting perspective has been a view of systems as interconnected wholes that have a coherent behavior at least potentially comprehensible to the human mind and intuition. Ecological science has mirrored this postmodern dichotomy. The ecosystem concept has been a vehicle for discussing and thinking about wholes in nature and society. The concept has been useful in environmental ethics, human ecology, ecological economics, and in biological conservation.

In this philosophical role, the ecosystem concept provides a way to interpret the idea of the whole, equilibrium, ecosystem evolution, and the organization of nature in space and time. I examine the bridging role of the ecosystem concept in these broader contexts.

Throughout the history of ecology, ecologists, and particularly systems ecologists, have insisted on studying whole objects in nature, asserting that "a whole is greater than the sum of its parts." A whole has emergent properties that characterize it.[10] Other ecologists disagree, insisting that the whole is merely the sum of the components that make up the system. Is there a way to deal with this issue?

Philosophers (Brennan, 1988, 86) require that wholes have genuine properties. A genuine property is one that is unique to the whole and not reducible to the properties of its components. Of course, the whole also has properties that are the sum of the properties of its components—but that is not the issue. Do ecosystems have genuine properties?

Let us consider two examples. First, I propose that the trophic structure is a genuine property of an ecosystem. The trophic structure is frequently described as a food web. Stuart Pimm (1982) begins his book on food webs with the comment, "Food webs are diagrams depicting which species in a community interact. They depict binary relationships—whether species interact or not—and must miss much important biology. In the real world species interactions change at least seasonally and not all interactions are equally strong. Food webs are thus caricatures of nature. Like caricatures, though their representation of nature is distorted, there is enough truth to permit a study of some of the features they represent." In his statement Pimm tells us that food webs depict the feeding relation between two individuals. This is the key point, and since cannibalism is reasonably rare (although it is not at all rare among certain taxa and habitats), one can reason analogically from the action of a single organism feeding on another to a single species feeding on another species. In this way, a trophic web may be constructed mentally, made up of species linked by feeding. This is the conventional food web of ecology. Nonetheless species is a category as troublesome as the ecosystem, and one could ask if the species is also a whole. One also could repeat Ivlev's question about variation in food habits among individuals making up the species. How do we categorize the herbivorous ground squirrel seen feeding on an animal carcass? Clearly, if the trophic web is a picture of the process of feeding of one organism on another, then the web is the sum of all of these feeding activities. Therefore, the food web is not a genuine property of the whole ecosystem; it is a property

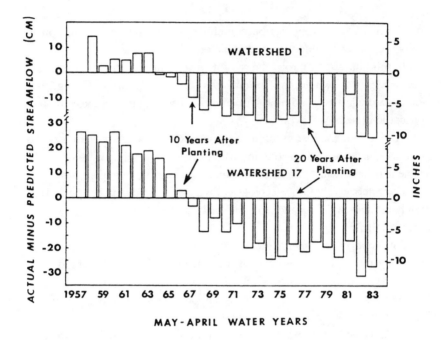

7.1 Annual changes in water flow from two watersheds at Coweeta Hydrological Laboratory, following the planting of white pine in formerly hardwood forest land (Swank and Crossley, 1988)

representing the sum of the feeding activities of individual organisms in the ecosystem.

Second, let us consider the flow of water and the chemicals carried by that water as sediment and dissolved materials as it exits a mountain watershed ecosystem at the Coweeta Hydrological Laboratory. Is the export of water and nutrients a property of the whole or is it the sum of the properties of the components of the ecosystem? The Coweeta basin is located in the southern Appalachian mountains of eastern North America and has been the site of studies on hydrology and chemical cycling for over fifty years. The normal vegetation on these mountain watersheds is a mixture of hardwood trees.

The Coweeta laboratory was established to determine the impacts from various treatments of the watershed on the stream flow (Swank and Crossley, 1988). Some of the hardwood forests were converted to a mountain farm and a white pine forest, others were cut using different logging methods in order to understand how these practices affected water yield and stream chemistry. For example, when the hardwood trees were replaced by white pine, stream flow

Vegetation Type	Chemical Ions								
	NO_3N	NH_4N	PO_4	Cl	K	Na	Ca	Mg	SO_4
White pine	−0.7	−0.1	0.0	+1.8	+1.7	+4.4	+2.3	+1.2	+0.7
Coppice	−2.2	−0.3	0.0	+1.0	+0.9	+2.8	+0.2	−0.6	+0.8
Grass-to-forest succession	−6.8	−0.1	0.0	−4.4	−0.7	−0.1	−3.3	−2.6	−1.0

7.1 Net loss or gain of selected chemical ions from three experimental watersheds at Coweeta Hydrological Laboratory compared with undisturbed hardwood forest watersheds (Swank and Crossley, 1988)

declined (fig. 7.1). The reason for this change is that pine trees remain in leaf throughout the year and the leaves continue to transpire water to the atmosphere. Thus, less water flows from the watershed as stream water under the pine forest, and more water is transpired to the atmosphere. The chemistry of the water under this treatment is also quite different. In general, the hardwood forest loses more nutrients in the stream water than the pine watershed (table 7.1). Is the export of nutrients a genuine property of an ecosystem? Yes, it is. Although each individual plant and animal on the watershed takes up and releases water and nutrients, there are in addition to these biological processes physical-chemical processes at work. Nutrients stored in the soil through chemical bonding are held in clay lattices. These nutrients are released chemically when hydrogen ions replace calcium ions on the binding sites. The outputs are not merely the sum of the component processes. The system output depends upon the interaction of the biota, the rock, the water, the atmosphere, and the soil. Thus, we can conclude that the water and chemical flux in the watershed ecosystem is a genuine property, and the watershed ecosystem is therefore, by definition, a whole.

This process of water and nutrient flux is not the only genuine property we can identify. Ecosystems also reflect solar energy from their surface (Ivorson, 1988; Ivorson, Graham, and Cook, 1989). The reflectance or albedo can be read by sensors in a satellite or airplane and used to determine the properties of the vegetation canopy. Reflectance can be correlated with physiological properties of the ecosystem, such as its primary productivity or evolution of water vapor or carbon dioxide, and the changing patterns of productivity of entire regions can be studied (Tucker, Townshend, and Goff, 1985). Reflectance of light is another genuine property of an ecosystem, since it depends upon the canopies of the plants, their physiological state, the water conditions of the site, the presence of insects or disease organisms, and so forth. The reflectance represents the interaction of all of these factors.

It is not necessary to cite the other genuine properties of ecosystems to be convinced that ecosystems are wholes, in the definitional sense proposed by philosophers. We note that genuine properties are not exclusively biological. Rather, they involve a mixture of biological and physical-chemical processes in the system. If we focus only on a biological process, the ecosystem property is often reducible to the biological individuals that are responsible for the action. This is not always true, however, which opens up another philosophical puzzle—that is, the property of symbiosis, mutualism, or cooperation. A mutualistic relation involves several biological entities acting together as a whole. For example, a family can be a whole. It is not merely the sum of its members; it has a historical, genetic, psychological, and cultural reality as well.

The establishment of the ecosystem as a whole is important from a scientific technical perspective. If it is a whole, then we can make statements about it based on our theoretical understanding of natural systems and test these statements with experiments. Though difficult to experiment with large objects like ecosystems, it has been done. The Hubbard Brook and Coweeta watershed treatments were experiments. The water yield from the watershed could be predicted from the history of water output over many years, what the hydrologist calls a *standardization period*. The reason that hydrologists need to standardize the watershed is that water yield depends in part on precipitation, which varies greatly from year to year and month to month. Thus, it is important to establish the annual and monthly variations in water yield before an experiment begins. Then, the vegetation on the watershed can be changed and water yield monitored for a number of further years. As we saw above, the prediction that water yield would be unaffected by a change from hardwood to white pines was not verified by the data. Indeed, the change in water yield was substantial, and the consequences of this change were seen in other chemical properties of the system. This type of experimentation was also a key feature of the Hubbard Brook study, and it has been a feature of post-IBP research at many of the sites of IBP biome studies.

Another well-known experiment, using an aquatic ecosystem, was carried out by David Schindler and his associates (1985) at the Experimental Lakes project in Canada. Again, these investigators established the normal variation in behavior of lakes in the region. Then, selected lakes were perturbed by adding chemicals to the water. For example, the pH of the lake water was reduced or increased by adding acid or lime. Following perturbation, the properties of the lake were monitored. Schindler and his associates found that ecosystem properties, such as productivity, were relatively robust and did not change under treatment nearly as much as did the roles of species in the system. Species dominance shifted under treatment. Rare species became common, and formerly common species became rarer. It appears there is considerable biological redundancy in these lakes, so the components shift their activity and abundance in response to changing environmental conditions, but the genuine properties of the lake, which are a function of the watershed and the atmosphere as well as the biota, are more robust and vary much less with an environmental change.

I conclude that the ability to see ecosystems as wholes requires a particular point of view. It is possible always to focus on the parts, to decompose every action or object into its components, and then to declare that all important properties reside in these components. David Bohm (1980) has criticized this tendency to see the world as fragmented into parts. He emphasizes that it is a point of view that not only characterizes much science but also the Western intellectual tradition and popular culture, and he calls for a change in our point of view and way of thinking. A capacity to consider both the parts and the whole is needed. Bohm is a spokesperson for the holistic postmodern perspective.

The holistic view is not the "right" view, it is merely another way of looking at the world. If we can agree that there are whole systems, with genuine properties, then we can investigate them with the reductionist scientific method and, within the limits of this method, discover useful things about them.

Equilibrium has played a special role in the development of the ecosystem concept. Not only did Tansley make it a central element of his ecosystem concept, but it reappears over and over through the history of the subject. It was one of the key elements of Stanchinskii's trophic ecology attacked by Prezent, who saw in equilibrium a natural constraint that could theoretically limit human control of nature. Stanchinskii responded to the Stalinists' attack by adopting a theory of randomness and chaos, anticipating the viewpoints of some Anglo-American ecologists by almost fifty years. His repudiation of equilibrium, however, did not save him.

What do we mean by equilibrium? First, what is the subject of our attention? Is it the genuine properties of the ecosystem? the secondary properties derived from the sum of the parts? or the property of a component such as a species population? Second, recognizing that the environment is continually variable and that biological organisms continually are born, die, adapt, and evolve, what limits of variance in space, time, and response do we accept within the concept of equilibrium?

In one of the few long-term studies of ecosystem performance, the Hubbard Brook study, it has been shown that system properties vary a great deal. Rather than equilibrium, this system is better described as a response system, that is in a dynamic relation with its environment. The state at any particular time is contingent upon its history and the environment. This is a relativistic theory of ecosystem behavior, and it contrasts strongly with Tansley's original deterministic concept. Ecosystems are loose systems, we could call them weak wholes, as compared to a strong whole such as an individual or a city. The changing environment in which the ecosystem is placed (that is, its landscape or biome) creates a dynamic response of the system as a whole. If the environment is changing in a consistent way, then the ecosystem will track that pattern. Further, the ecosystem may have a damping and controlling influence on the environment; it has a reciprocal relation to its environment and is not merely responding to it.

If it requires ten to fifty years of monitoring an ecosystem's properties to determine the equilibrium conditions of the ecosystem, we could consider equilibrium to be the history of the system. It is a statement about its past performance, not a prediction about its future state. Can we predict the future state? No, but we can describe broad limits of probability. A hurricane may destroy a forest by overturning many trees, acid rain may gradually poison a forest, an insect outbreak may defoliate the trees, economic forces or political pressure may induce forest managers to cut trees, and so on. Yet if the likelihood of these events occurring is small, then we can predict the water flow and chemical export from the watershed within broad confidence limits. This is a technical explanation for the popular conception of balance in nature.

A more modern approach to the existence of equilibrium is through the concepts of resistance and resiliency, which describe how an ecosystem might respond to or recover from a disturbance. Resistance to disturbance is a combination of primary and secondary properties of the system. The structural mass of the biota, the capacity to store essential resources, the redundancy of essential components (usually expressed as high species diversity), a history of survival of past disturbances, tight control by the physical environment, operative feedback processes, and the capacity to shunt most of its productivity to maintenance processes all contribute to resistance. Resistance usually does not require all these properties to operate at the same time. Eventually, the biota evolves to fit a certain sequence of environmental disturbances, thereby enhancing the systems capacity to resist. For example, Ariel Lugo (1990) and colleagues (Lugo, Sell, and Snedaker, 1976) have shown that mangrove forests growing along the coast of Florida experience a severe hurricane about once in twenty years. The growth and reproduction pattern of the mangrove trees fits this temporal sequence, so that the mangrove ecosystem can persist in that environment.

The capacity of an ecosystem to respond after being disturbed is called resiliency. Resilience is a function of the scale and intensity of the disturbance, the presence or absence of the biota, isolation, the presence of organic remains in the soil, soil fertility and water-holding capacity, the presence of toxic substances, and so on. If the system has been completely removed from a site, then the capacity to respond might be severely limited and recovery may never return the system to anything resembling the original state.

It is only in erratic environments, such as one created by a volcanic eruption or one dominated by modern humans, where the ecosystems are disrupted so frequently or so fundamentally that natural systems are chaotic and unpredictable. The deterioration of the amphibian fauna, a loss of bird species, a dieback of forests, and an increasing sickness of children and elderly are all symptoms of chronic ecosystem illness. This syndrome of disturbance has reached extreme conditions in some countries (Ryszkowski and Balazy, 1988), and it is increasingly observed in other parts of the biosphere. Under extreme conditions of stress it is possible that ecosystems will behave chaotically.

The word *evolution* was applied to ecosystems in the early 1970s (Darnell, 1970; Cloud, 1974). It was an unfortunate application of the term. Evolution has a distinct scientific meaning relating to genetic variation, natural selection, and the formation of new genotypes and phenotypes. Since ecosystems have no genetic structure or even an analogous genetic structure, there can be no ecosystem evolution in the same way that there is evolution of a biological species. Still, ecosystems do change over time, and we can speak of ecosystem development. The developmental pattern of ecosystem structure and function has two aspects. First, at the broadest scale of time, in millennia, ecosystems have changed in form and distribution over the earth's surface. Second, within shorter time periods, ecosystems undergo ecological succession. Both types of temporal change have led to theoretical developments that have shaped the ecosystem concept.

Paleoecological work by Margaret Davis (1981), Hazel Delcourt, and Paul Delcourt (Delcourt and Delcourt, 1988) in eastern North America have shown that the deciduous forest that covered the continent at the time of European invasion did not have the same complement of species that it did in the Pleistocene era. Ecologists had thought that the great biomes of tundra, boreal forest, and deciduous forest remained relatively intact and shifted south with glaciation in bands across the continent and then shifted north again as the glaciers receded. The evidence shows, however, that species moved individually at very different rates of speed. Some species survived and were common during glaciation, while other species became less common and disappeared. In other words, the taxa making up the flora, as revealed by pollen analysis, changed in abundance and distribution so that the species composition of the biomes changed over time. Since we know that an ecosystem is created by the interaction of the biota and the physical-chemical environment, and that this interaction creates unique properties such as soil and the gradients of carbon dioxide, hydrocarbons, and oxygen in the ecosystem atmosphere, we can speculate that the deciduous forest biome of the past probably did not function in the same way as it does today. But we do not know if our speculation is true. The point is that paleoecology has shown that ecosystems are not fixed types that are constant over long time periods. Rather, they are adapting and changing as the individuals within them adapt and change. Ecosystem development is a function of the success and failure of its biotic components.

On a shorter time scale, ecologists have studied and speculated about ecological succession endlessly. This topic forms a major theme in ecosystem studies. In the new form of ecosystem theory, succession is considered as an aspect of resilience of the system, subjected to disturbance. Successional theory is being reinterpreted in this new form.

In considering the role of the ecosystem concept in understanding temporal and spatial scale, it is helpful to return to a theme dealt with earlier. Arthur Koestler (1978) adopted the Roman God Janus as a metaphor to represent how we approach a scientific question. Janus had two faces that looked in opposite directions. Koestler interpreted this metaphor as a bidirectional strategy, which we follow after discovering a pattern in nature. We ask two kinds of questions. First, why does this pattern work in this particular way? To answer the question requires us to look inside the system, identify the relevant components and discover their behavior, experiment with them, and then determine how each contributes to the system patterns we discover. I labeled this strategy *reductionistic* in my earlier discussion. Second, what is the significance of the pattern? How does the behavior of the system contribute to larger patterns of behavior in systems of which our system of interest is a part? We can call this strategy *synthetic*. In pursuing it, we look outside the system of interest and attempt to see how it is connected to others, how it is influenced by others and reciprocally influences them, and what its role is in the larger world. If we are able to complete both analyses, then we have a more complete study and can apply our knowledge with more confidence. We also have information to erect hierarchies of scale that describe the connections between the levels of relationship.

Ecosystem studies have followed this line of development, but the effort in two directions has not been simultaneous or balanced. First, it was necessary to establish patterns of ecosystem function. The Lindeman-Hutchinson analyses were a first attempt to do this. Their work was followed by many other individuals and then was greatly expanded by the IBP biome studies. The accumulating database has shown that ecosystems vary widely in patterns and rate processes. Although certain patterns—such as the rates of organic productivity of vegetation and organic matter decomposition—are relatively predictable, others (such as the function of specific animals) are not. The hope that ecosystem processes would converge on specific efficiencies and rates has never been realized.

Second, ecosystems were not always studied reductionistically. Ecosystem studies did not always proceed step by step from the discovery of a pattern to identification of those components relevant to that pattern, and then to experimental or observational research on those components. The concept of the ecosystem was being developed by biologists who were building from studies of populations and communities. This meant that some of the internal analyses of the biota were in place before the ecosystem pattern was established, and they strongly biased the work. For example, ecologists debated the way in which groups of organisms were to be constructed. Were grass-eating animals that ate 10 percent insect food to be included in the herbivore compartment or not? This question was relevant only if construction of food categories was essential in answering questions about ecosystem function. Rather than decomposing the ecosystem based on the hypotheses proposed, sometimes the process of concept development was backward, and the system was constructed without knowing the assembly rules. This confusion created many problems of understanding and bitter quarrels.

Further, initially there was no clear concept of a larger system of which the ecosystem was a part. There were words such as *ecosphere, biosphere, biome, region,* and *landscape,* which indicated that organized systems existed at large spatial scales, but ecologists did not explore this scale in the beginning of ecosystem study. Why? Probably the main reason was the lack of remote-sensed data. The large spatial perspective could only come from areal photography, and the size of photographs was too small and the collection of data from the photos was tedious and difficult. Remote sensing and computer processing of remote sensed data changed the situation dramatically, and we now have a variety of ways of seeing large surfaces, from the earth as a planet down to a single square kilometer. In addition, ecologists did not frequently interact with geographers, whose discipline is concerned with large regions and with scale problems. The consequence was that it was difficult to develop Arthur Hasler's and Eugene Odum's proposal that the IBP study whole watersheds and include both the land and water within a single system, although Glenn Goff in the Eastern Deciduous Forest Biome project tried to take a regional approach, and the Lake Mendota group at the University of Wisconsin also had this broader perspective. Finally, ecosystem studies focused on a single plot, site, or measurement area judged to be representative of the larger system. Little attempt was made to replicate studies or establish the central tendencies of ecosystem behavior. Ecosystem studies, until the current phase, seldom had dimensions in space or time (Allen and Starr, 1982; O'Neill et al., 1986).

Since research funding is limited to short periods, especially in the United States, it has been exceptionally difficult to study a system over adequately long periods. Ecologists recognized these limits as seriously affecting their work. It is not clear whether the IBP and Hubbard Brook results revealed the problem, or whether these experiences gave ecologists who were aware of the problem the confidence to organize and try to change the system of funding. In any event, after the IBP ended, an effort was organized to develop long-term ecological research. "Long-term" in this case meant funding for five years, with the likelihood of further support. Eventually, these efforts were successful, and in 1979 a competition was held by NSF to select six sites for long-term ecological research funding. During the next ten years the number of sites was expanded to almost twenty, and with these funds for the first time long-term ecosystem research was possible. This dimension is absolutely necessary for a knowledge of ecosystem performance, especially if the model of ecosystems as response systems is accepted.

Therefore, by about 1990 there was in the United States a system for funding ecosystem research at proper spatial and temporal scales. It required about forty years, or two generations of ecologists, to put the organizational system together, institutionalize it properly, and arrange dependable funding.

THE CHALLENGE

The examination of the ecosystem from a conceptual perspective indicates that the concept has a variety of meanings. As a philosophical idea, it animates research and application in environmental studies and management. As a concept that identifies a physical object (in Tansley's sense), it is the subject of much scientific study. As a scientific paradigm, it structures scientific organization and research. With such broad usage, it is no wonder that the ecosystem concept has been challenged by many ecologists. It is essential that we consider these challenges in order to complete this evaluation. I am not going to consider criticisms about the details of ecosystem studies, such as the way population aggregations have been done, or attempt to answer the criticisms or comments of specific individuals here. Ecosystem scientists have made many peculiar statements that are sometimes contradictory, just as with any group of individuals. Rather, I am concerned with the broader challenges to the ecosystem concept.

The Superorganism

The ecosystem concept has been criticized as a continuation of the superorganism idea espoused by Clements, and it has been dismissed as organic holism. The superorganism was a metaphor that attributed the properties of organisms to vegetation (or the biome) as a whole. It was appropriate to a time when organic metaphor was popular and widely employed by technical and popular writers. It is a misleading metaphor because an organism and vegetation have no points in common, even when one considers the process of development. One problem with this form of reasoning is that the plant community was made an abstraction—distinct from concrete stands and patches of living plants—and a universal idea. It was then possible to attach the abstraction to idealistic concepts such as those of John Phillips.

This was the issue Tansley addressed when he formulated the ecosystem concept as an alternative to superorganisms and similar general categories. That ecologists still confuse Clements's superorganism and ecosystems testifies to the persistence of concepts in ecology. In a science that is increasingly relative, faced with countless objects, and with little tested theory, these generalizations may be life rings that we hold onto in the absence of something more reliable. With a fully operative hierarchical concept in place, there is no longer any need to create general categories to hold vegetation, communities, or ecosystems. Landscapes hold ecosystems, and one can study a landscape as an entity just as one studies any other unit of the natural world. It is ironic that the ecosystem concept has been criticized for being superorganismic when it was proposed as an alternative to the superorganism.

Further, a problem of scientific method has been raised. Richard Yeo (1986, 277), discussing the scientific method in Britain from 1830 to 1917, showed a shift from the Baconian inductive approach based on generalizations from a collection of observations to the hypothetical-deductive approach. Yeo pointed out that in the latter half of the nineteenth century there was considerable contention between the physical and organic sciences. Zoology, botany, and natural history were placed in a subordinate position with regard to physics, which was "presented as the model to which all sciences should aspire."

Sciences have demonstrated their "physics-like" nature by emphasizing their use of the hypothetical-deductive method. In ecology, the debate between proponents of the Baconian and hypothetical-deductive methods has been continuous. In the 1970s, especially in the United States, there was a strong attempt to drive Baconian methods of science from ecology. The effort was not entirely successful, partly because it is difficult to experiment with ecological systems and partly because of the contingency of system behavior on individuals and chance.

Ecosystem studies share these characteristics with other sciences, such as geology, atmospheric science, and ocean science. The development of ecosystem modeling provided a tool for improved induction, but it remained Baconian. This division in ecology has had an unfortunate impact on the role of ecology in environmental management. Eugene Hargrove (1989), in his book on environmental ethics, pointed out that the preferred hypothetical-deductive model of scientific inquiry "encouraged scientists not to think about the environment as it is encountered in experience," thereby inhibiting environmental concern. These arguments about methods may have made ecologists less effective in solving environmental problems.

Determinism

Another criticism of ecosystem studies, advanced by Daniel Simberloff, is that ecosystems are deterministic. That is, that ecosystems are goal-directed and self-regulated. In contrast. Simberloff thinks that ecological systems are stochastic and indeterminate. Indeed, he interprets the history of science as a dialectical opposition between determinism and indeterminism.

Simberloff is correct when he says that certain prominent ecosystem ecologists have asserted that ecosystems are self-regulated, goal-seeking systems. Both Bernard Patten and Eugene Odum (1981) made such statements, as have others. Be that as it may, other ecosystem scientists have described the stochasticity of ecosystem behavior. For example, Bormann and Likens (1979) described the variation in chemical export from the Hubbard Brook watershed over many years and have shown how the system tracks the environment but is influenced by its history, its physical-chemical character, and the internal dynamics of its populations. David Schindler and his associates (1985) have shown through experimental manipulation of whole lakes how properties of the lake, such as productivity, remain within limits of variation while the species populations that dominate the producers and consumers change.

Thus, it is not true that the ecosystem approach requires that ecosystems function deterministically. Rather, it is likely that ecosystems evidence probabilistic behavior over space and time, although this is less true where physical or chemical constraints tightly shape the possible biological responses. For example, on tundra or in deserts cold temperature or lack of water may permit only a few biological strategies to be employed, so that there is something of an either-or situation. Further, processes tightly coupled to energy from the sun tend to be more tightly constrained. Where the biota can be active and diverse, the individual genetic differences and the potential variety of adaptive responses to environment create a probabilistic situation, and the variation in response is as important as the modal response. Stochasticity can be programmed into an ecosystem model, so this insight can influence the forms of potential predictions as well.

The Ecosystem Is Cybernetic

Joseph Engelberg and L. L. Boyarsky from the department of physiology and biophysics of the University of Kentucky in 1979 published an article claiming that ecosystems were not cybernetic systems. This claim struck at the heart of a tenet of ecosystem studies that stated that ecosystems were regulated through feedback processes. Engelberg and Boyarsky classified systems into cybernetic and noncybernetic and defined cybernetic systems as characterized by information networks linking all parts of the system together. Linkage was by feedback loops channeling information and, as a consequence of these information networks and feedback loops, the system was stabile. Examples of cybernetic systems, according to these authors, are the human organism and cells. Ecosystems were classed as noncybernetic because the authors denied that ecosystem components are connected by an information network. Rather, they argued that the dominant exchange between components is informationally nonspecific matter and energy.

This is a curious criticism, which we might ordinarily ignore as an individual case interesting only in itself; however, it raises an important point about ecological networks. If one focuses on components or individual organisms as a biologist would tend to do, you have a very different systems perspective than if you focus on the flows of energy, matter, and information directly. Systems models tend to be drawn so that components or sets of components are linked together in a network. Actually, the ecosystem consists of coevolved suites of organisms. Carl Jordan (1981), in response to the Engelberg-Boyarsky article, cited Larry Gilbert's *Heliconia-Passiflora-Anguria/Gurania* complex of tropical forests as an example of coevolution. Further, as Robert Paine (1969) pointed out, there are keystone species that provide special environments for many other groups. There also are social organisms, such as ants, that form yet another pattern of organization. This means that the actual organization of an ecosystem is much more complex than the network model suggests. Indeed, the organization of a large city might be a better model than the systems models of textbooks, the links of which if very complicated look like a bowl of spaghetti.

It does not seem useful to divide systems into cybernetic and noncybernetic systems, because this classification brings us back to the point of Simberloff and the typology of deterministic systems. Further, it has not been clear in ecology what is meant by information. Rather the question is, how is control mediated in ecosystems? An example of a potentially complex control network has been the focus of a decade of research by Melvin Dyer (1980) and Jim Detling (Detling et al., 1980), working at Colorado State University and elsewhere. These ecologists have proposed that bison, grasshoppers, and other herbivores on the grassland not only negatively impact plants by the removal of leaves when feeding and by their trampling but also positively influence plant growth through their manure and the process of feeding. They have proposed that chemical compounds in the animals' saliva or from mouthparts may stimulate plant growth as a response to feeding. This could be a feedback system whereby the plant-herbivore dyad is linked in both positive and negative information flows.

These challenges to the ecosystem concept focus on the nature of the ecosystem and its properties. They tend to place the ecosystem concept in the past, appropriate to a stage of ecology that has past, and imply that it has been replaced by new forms of ecology, such as evolutionary ecology. This implied contrast is a major theme in the study of the ecosystem concept by Joel Hagen (1992). Frankly, I feel that organizing ecology around a conflict theory does not adequately describe the complexity of ecological research. Many individual ecologists work in population, community, and ecosystem ecology analysis at the same time. Others have moved in their careers to deeper levels of analysis of a phenomenon and therefore focus on levels of organization finer than the ecosystem. The elements that are used to place the ecosystem concept in the past are its connection to deterministic, organic wholes and the tendency of some ecosystem scientists to employ Baconian science rather than hypothetical-deductive methods. When ecosystem scientists propose that ecosystems are self-regulated superorganisms with the purpose of maintaining stability, these criticisms are justified. The criticisms are not justified when we consider the mature form of ecosystem studies, which considers flows of energy and cycles of materials in well-defined systems. It is this modern ecosystem approach that provides the basis for the study of global change and other biospheric problems.

If we follow the ideas of Heraclitus, rather than the atomism of Democratus or the concept of ideal types of Plato, and recognize that all nature is ever-changing, then we realize that the ecosystem concept is merely one more device for thinking about the world and ourselves. In the Heraclitian interpretation, the real world "out there" actually consists of fields of energy, matter, and information, which we stop in thought and language as if we were taking a snapshot of reality. The methods we use tend to enhance our capacity to see these "objects," whether we use a microscope that lets us see into a cell or a telescope that enables us to see into the universe. But we do not have equally powerful methods to see flux. What would our worldview be if we could do so? Intuitively, the field ecologist senses flux in nature and sometimes uses the ecosystem concept to express this sense of dynamism. How can it be made visible to others?

Ecologists were projected into prominence in the 1960s with the public's recognition of the environmental crisis. The prominence was not sought and was rejected by many. In the process of popularization the word *ecology* became a synonym of environment, and *ecologist* became a term adopted by anyone concerned about environmental deterioration. Some of the individuals involved in this story became deeply involved in environmentalism. There was, it seems, some connection between their interest in the ecosystem concept and their environmental concern. The two interests fed back on one another. The technical ecosystem concept developed by these scientists was interpreted to lecture audiences and in popular articles as a commonsense, natural wisdom with an up-to-date system, cybernetic, computer-like character. Obviously, popular ecosystem science simplified and thereby, in a sense, misrepresented the technical science. Still, it also captured the interest of a diverse group of environmentalists and environmental practitioners. Further, the emphasis on relation, connectedness, and dynamic but balanced function in a holistic context linked this popular machine metaphor to the age-old interest in whole systems, where the whole becomes more than the sum of the parts. The concept of the ecosystem in this form has stimulated philosophers (Callicott, 1990) to think more deeply about the nature of nature and human relations with nature. Ecosystem studies has contributed to a new philosophical subject called *environmental ethics* or *ecological philosophy,* which has become a dialogue about how humans value nature (Golley, 1987; Cahn, 1988). It is not clear to me where ecology ends and the study of the ethics of nature begins, nor is it clear to me where biological ecology ends and human ecology begins. These divisions become less and less useful. Clearly, the ecosystem, for some at least, has provided a basis for moving beyond strictly scientific questions to deeper questions of how humans should live with each other and the environment. In that sense, the ecosystem concept continues to grow and develop as it serves a larger purpose.

NOTES

Chapter I: Introduction

I In hearings before the Subcommittee on Science, Research, and Development of
the Committee on Science and Astronautics, U.S. House of Representatives,
90th Congress, 9 May, 6 June, 12 July, 3 and 9 August 1967, Frederick Smith,
director of the U.S. IBP biome program stated, "An ecosystem is a physical entity
which can be studied" (p. 23), and, "I should point out that nobody has ever
studied whole ecosystems. It has never been done" (p. 24).

2 Ernst Haeckel, the famous German interpreter and defender of Darwin, presented the term *ecology* in a text titled *Generelle Morphologie der Organismen*. An
English-language translation of the paragraph defining ecology appeared as the
frontispiece in Allee et al., *Principles of Animal Ecology*: "By ecology we mean the
body of knowledge concerning the economy of nature—the investigation of the
total relations of the animal both to its inorganic and its organic environment;
including above all, its friendly and inimical relations with those animals and
plants with which it comes directly and indirectly into contact—in a word, ecology is the study of all those complex interrelations referred to by Darwin as the
conditions of the struggle for existence. This science of ecology, often inac-
curately referred to as 'biology' in a narrow sense, has thus far formed the
principle component of what is commonly referred to as 'Natural History.'"

3 I have used the quotation of John Harper and the comments about V. C. Wynne-
Edwards from Kimler (1986, 232). The quotation is from Harper, 1977b, 148.

4 The Association for Ecosystem Research Centers was organized in about 1985.

Chapter 2: Genesis of a Concept

1   The Tansley biographical material includes Sir Harry Godwin's three publications, two of which concern Tansley directly (Godwin, 1957, 1977) and the other (Godwin, 1985) with Cambridge, England, and includes a discussion of Tansley in this context. G. Clifford Evans, in his 1975 presidential address to the British Ecological Society, explored the development of Tansley's ideas, including the ecosystem concept. In addition, Tansley presented several analyses of the ecological sciences of his own, including a review of British ecology over the past quarter-century (1939b), the value of science to humanity (1942), the early history of modern plant ecology in Britain (1947), and *Mind and Life* (1952).

Few of Tansley's papers remain, although Sir Harry Godwin, responding to a letter from R. E. W. Maddison, The Royal Society, about Tansley's papers, said that he saw no files of correspondence, notebooks, or manuscripts when he went through Tansley's office after his death, and as a consequence, that he thought these materials were small in volume or had been disposed of by the family. The primary collection is in the Botany School, Cambridge, along with Tansley's reprint collection. In addition, there are several important letters of Tansley in the Frederic Clements Collection, American Heritage Center, University of Wyoming, Laramie (hereafter referred to as the Clements Collection). Tansley's library was inherited by Godwin but was disposed of upon Godwin's death. Peter Grubb, Cambridge, owns a small set of Tansley's ecological volumes.

2   A letter, dated 18 Dec. 1918, from Tansley to Frederic Clements referring to a quotation in German from Professor Gams, Innsbruck, about Clements, said that Gams was "very angry indeed! Almost as angry as Professor Bower of Glasgow on the subject of my 'Bolshevism.' I've been getting some experience in the 'Gentle art of making enemies' lately. The more you keep your temper the madder they get. Reactionary forces are pretty strong here, and it will be a hard struggle to get anything progressive done. But I am going to have a good try. In regard to the 'reconstruction' discussion the enemy has had his innings and the end of it will be mainly on my side. But it is a long step forward to deeds, especially when the high places are occupied by the enemy. Fortunately my livelihood does not depend upon the favor of the exalted reactionaries. But I often look with envy on your 'hope-filled western skies' (Chicago *Alma Mater* hymn).

Nevertheless, my job is on this side. I am sure though it is sometimes depressing to realize that now the Boche is beaten we have to begin another fight in the spiritual sphere." Letter, Clements Collection.

Clements's response to Tansley, 14 Feb. 1919, was: "I was greatly interested in Bower's article about the newer teaching. Naturally you are in a difficult position with the most important chairs occupied by men of that type. It is marvelous how thoroughly hide-bound static subjects such as morphology can make a man. Or perhaps it is merely that everything tends toward stabilization and nothing but the extremist devotion to progress can prevent it either in the individual or his work." Letter, Tansley Papers, Botany School, Cambridge.

Finally, in a letter to Clements, 12 July 1923, Tansley described his relief to be free of the Cambridge conflict. "I am most thankful to be free of obligations as University Lecturer—I scarcely realized until I had actually resigned what a strain being part of an uncongenial, uncorrelated organism like the Botany School, Cambridge, really was." Letter, Clements Collection.

3 On 12 Jan. 1923 Clements wrote Tansley asking about his decision to resign from botany: "I have been hoping to hear from you with reference to your decision and plans for some number of the Journal of Ecology, I can understand why you have no time to spare for letters. However, I am anxious to know what you are planning to do, and still hope that you have been able to arrange matters so that you will not have to forsake ecology altogether. Perhaps it is selfish on my part because I am not at all sure that your new field may not have greater opportunities for distinct and distinguished services." Letter, Tansley Papers, Botany School, Cambridge.

On 8 March 1923 Tansley answered, "Probably I shall cease to be a professional botanist after the term, though for the present, at least, I shall continue to edit the two journals. . . . Admonson is going to the Cape and will be a terrible loss to me—I need a good 'florist' at my elbow. Together with the 'conservatives in authority' his departure will help make me spend more time at psychology and less at ecology. The last year or two I have been pursuing both, and though my power of work is much better than it was, largely I think to the release of powers through emotional clarification—the double pull is a considerable strain." Letter, Clements Collection

Then, on 30 May 1923, in another letter to Clements, Tansley wrote, "You will be interested to hear that I have now definitely resigned my University Lectureship in Botany. I am tired of official lecturing and I do not see the possibility of doing anything better in the teaching line within the existing framework, which I can not alter. I go to Freud again in October for some months, but for the present, at least, I shall continue to edit the two journals. It is likely that I shall take my whole family with me to Vienna." Letter, Clements Collection.

4 Tansley, 1935, 284. In the second decade of the twentieth century, Frederic Clements published his monumental book on plant succession and became the individual most closely associated with the topic. By the time of the Cowles festschrift, Tansley had distanced himself from Clements extreme interpretations. Their difference of opinion, however, had been apparent as early as 1915 and 1916 (see letters, Clements Collection). Clements expressed the hope that they could come to an agreement over the meaning of the concept of habitat and formation. In Tansley's own discussion of succession (1929), he analyzed the concept in a broad and advanced way, anticipating many of the views of later antagonists of the Clementsian concept. For example, Tansley states that "a climax community is a particular aggregation which lasts, in its main features, and is not replaced by another, for a certain length of time; it is indispensable as a conception, but viewed from another standpoint it is a mere aggregation of plants on some of whose qualities as an aggregation we find it useful to insist. . . . These selective syntheses are essential to the progress of science, and the particular ones

mentioned are of very great value, as I have tried to show, in the study of vegetation and ecology. But we must never deceive ourselves into believing that they
are anything but abstractions which we make for our own use, partial syntheses
of partial validity, never covering *all* the phenomena, but always capable of im-
provement and modification, preeminently useful because they direct our
attention to the means of discovering connections we should otherwise have
missed, and thus enable us to penetrate more deeply into the web of natural
causation." "It is the special credit of American ecology, and in the first place of
the labors of Cowles and Clements, followed by a host of gifted workers, that
laid stress upon the successional way of viewing vegetation at a critical epoch in
the development of the science" (p. 686).

Tansley's view of succession was more balanced and more empirical than
Clements's. In this sense he is closer to Cowles—who described the vegetation of
the Lake Michigan dunes and interpreted it within conventional plant ecology—
than to Clements, who created a new ecological paradigm based on his observations of vegetation change.

5 Tansley asked Clements about Phillips in two different letters. On 17 July 1924,
he asked "Who Is Phillips?" and on 12 October 1924, "I am curious that I
haven't heard of Phillips. One would have expected he would have contributed
to the perambulations and discussions we had on Empire vegetation work at the
last Imperial Botanical conference in July. I wonder if the secretary missed him,
so that he never heard of it?" (Clements Collection).

Phillips, in a tribute to Clements published in *Ecology* in 1954, described his
own academic history. Phillips was educated at Edinburgh but moved to South
Africa in 1922. He remained in Africa throughout his career and was a professor
of botany at the University of Witwatersrand from 1931 to 1948.

6 Phillips, 1931, 20. The references to Clements are 1905, 1916, and Clements,
Weaver, and Hanson, 1922; to Tansley, 1920 and 1929; to Smuts, 1926.

7 Tansley, 1935, 285. Phillips, however, wrote about thirty years after this article
(in 1954): "Tansley, in this journal in 1935 (16:284–307), in a kindly manner,
hinted that my papers in the Journal of Ecology (1934–35:22, 23) on succession, development, the climax and the complex organism suggested a tendency to
absorb the pure milk of the Clementsian word. I still hope to publish a corrective
to my old friends' courteously incorrect assessment." Even so, Phillips's last comment on Clements was adulatory and did not deal directly with Clements's basic
concepts.

8 In a letter, Clements to Tansley, 6 Dec. 1916, Clements writes, "I fully appreciate
the great advantage we have over here with our enormous stretches of fairly uniform untouched climaxes" (Tansley Papers, Botany School, Cambridge
University).

9 The term *holism* was derived by Smuts from the Greek *holos,* or "whole", and the
English suffix -*ism*. There is a deeper origin for this English word, however.
Barnhardt (1988, 1229, 1234) states that the original spelling of whole was *hol,*

which was derived from Old English *hal*. The *wh-* spelling for words beginning in *ho-* began to appear in the 1400s. Thus, Smuts was doubly correct is choosing his spelling over the widely used *wholism*.

10 An élan vital, Henri Bergson (1911).

11 Wheelwright (1959) interprets the relevant fragments of Heraclitus's writings that have survived as follows: "Everything flows and nothing abides; everything gives way and nothing stays fixed. You cannot step twice into the same river; for other waters are continually flowing on" (p. 29). Opposition brings concord. "Out of discord comes the fairest harmony" (p. 90). "And, it is wise to acknowledge that all things are one" (p. 102).

Chapter 3: The Lake as a Microcosm

1 The term *limnology* was coined by Forel from the Greek *limne* or lake (Rodhe, 1974, 67). Einar Naumann and August Thienemann (1922), however, in proposing the formation of an international association of theoretical and applied limnology wrote: "Limnology is the science of fresh water as a whole, and includes everything that affects fresh water. It falls therefore into two parts, hydrography and biology." Hydrography includes the study of the form of the lake basin, deposition processes, physics of water, temperature patterns, water chemistry, and so on.

2 I am indebted to Sharon Kingsland (1985) for her information on Forbes and for pointing out his use of the ideas of Spencer and Darwin. Spencer is one source of the concept that communities are organisms. He stated that societies of humans are organisms and drew analogies between the development of the individual and development of society. Indeed, he used the phrase "structure, function, and development of the system" in referring to social systems (Andreski, 1971) . The language of Spencer is almost exactly that used by ecologists in discussing ecosystems.

3 Ward and Whipple (1918) used the term *society* to refer to the organisms living in specific habitats. For example, there were limnetic societies and littoral societies. Limnetic societies were divided into "placton" (sic) and "necton" (sic). There were also lentic societies in stillwater and lotic societies in flowing water.

4 In Forbes's (1907) words: "By 1879 . . . a virtually new situation had arisen in science, and especially in scientific education. Under the influence of Darwin and Agassiz and Huxley, a transforming wave of progress was sweeping through college and school, a wave whose strong upward surge was a joy to those fortunate enough to ride on its crest, but which smothered miserably many an unfortunate whose feet were mired in marsh mud. This wave reached central Illinois in the early seventies" (p. 895).

5 Thienemann, 1925, 20–22. Trans. F. B. Golley.

6 Harald Sioli, an emeritus director of the Max Planck Institute for Limnology at Plon, Germany, stated (in a letter to the author, 12 Feb. 1988): "Thienemann, in

every case, did not use the term [meaning the word *ecosystem*]. I have never heard it in his, or in Professor Lenz's lecturers at the Kiel University where I studied 1931–1934. Instead, Thienemann liked to speak of what we now call ecosystem as an 'Organismus hoherer Ordnung' [that is, as an organism of higher order] which comprises the biotope and biocoenosis in interaction and as a unit." Gerhard Trommer, in a conversation, said that Thienemann had actually used the term *biosystem* for this idea, as well. The connection between biosystem and eco-system is obvious.

7 References to these philosophers and biologists are Smuts (1926); Meyer-Abich (1938, 1948); Friederichs (1937); Weber (1939a and 1939b); and Woltereck (1940).

8 Salm, 1971, 11, comments that "Goethe practiced, described and continuously urged a radically different approach to nature, in which the phenomena was not to be broken down into its individual components but perceived instead in its to-tality."

9 Möbius used several spellings for his new term, biocoenosis. The word was co-ined from the Greek words *bios,* meaning life, and *cenosis,* meaning being together. *Biocoenosis,* in the German spelling, then, meant living organisms being together in one place, that is, in an ecological community. The English spelling is usually *biocenosis,* and I have tried to use it throughout, except in the original Möbius translation and where authors used an alternate spelling.

10 According to Croker, 1991, 35–37, Elton was influenced by Victor Shelford's book *Animal Communities in Temperate America,* which organized the enormous amount of data collected by ecologists on the animals of a community into food relations. The study of food relations represents what Elton termed the *food chain*—a more effective metaphor. Elton (1966) said that he carried Shelford's book in his knapsack when he made his first trip to Spitsbergen.

11 Cooper's major studies of ecological succession were made at Glacier Bay, Alaska, where he studied plant development of exposed surfaces after retreat of the glaciers (Cooper, 1923).

12 The Yale University Library Manuscripts and Archives Collection holds in the Raymond Laurel Lindeman Collection several boxes of his notebooks. In the productivity notebook, Lindeman recorded analyses of the fat, protein, and car-bohydrates of ecological samples from Cedar Bog Lake. He then multiplied these by their caloric equivalents and summed the values to obtain the energy stored in the sample. It appears that he chose Birge and Juday's data over these direct mea-surements for his energy calculations.

13 Deevey discussed efficiency in a 31 Oct. 1940 letter to Lindeman (Raymond Laurel Lindeman Collection, Yale Univ.), calculating it as: primary con-sumers/producers [times] X = secondary consumers/producers. [The actual numbers were 2.08/121 [times] X = 1.17/121 = .563. . . .] The actual numbers were 2.08/121 [times] X = 1.17/121 = .563. Deevey concluded that this result was too high because of feeding by other organisms. It was necessary to allow for this and for the loss to metabolism. Therefore, he recalculated efficiency as: 2.08 + 1.17/121 X = 1.17/121 = X = .361. Deevey concludes by commenting, "This idea was Evelyn's—I confess I don't thoroughly understand it."

14 It is not clear to me what material Hutchinson contributed to Lindeman's article. Hutchinson's letter to Park, 18 Nov. 1941 (Yale University Manuscripts and Archives Collection) in response to the rejection of the trophic-dynamic article, stated that "most of the *specific* points challenged are matters for which I, rather than Lindeman, are responsible." Further, he indicated that much of the material at the end of the article had occurred to him independently. We also have Lindeman's letters to colleagues and mentors in which he repeatedly comments on Hutchinson's help and contributions. Hutchinson was concerned, however, that Lindeman's reputation not suffer because of his pressure on Lindeman to publish the theoretical trophic-dynamic article. It is not certain if Hutchinson was taking on more responsibility to protect Lindeman or actually had made substantial contributions to the article.

15 The letters of rejection are in the Yale University Manuscript and Archives Collection. Comments of reviewer no. 1 (who was probably Chancey Juday) included the following comment: "A large percentage of the following discussion and argument is based on belief, probability, possibility, assumption and imaginary lakes rather than on *actual* observation and data. . . . According to our experiences, lakes are *rank individuals* and are *very stubborn* about fitting into mathematical formulae and artificial schemes proposed by man." Reviewer no. 2 (who was probably Paul Welch) commented: "The paper is an essay and papers in Ecology should be research papers. This kind of treatment is premature. . . . Limnology is not yet ready for generalizations of this kind."

16 See note 14, this chapter.

17 Hutchinson (1979, 246–48) reflects on this difficult time in his autobiography: "A deep suspicion of theoretical formulations was probably most marked among the biologists of the middle-western states, where plant ecology was rapidly growing. It came very striking to my attention later in my career, when I was attempting to get Raymond Lindeman's famous paper, 'The Trophic Dynamic Aspect of Ecology' published. This paper was the first one to indicate how biological communities could be expressed as networks or channels through which energy is flowing and being dissipated, just as would be the case with electricity flowing through a network of conductors. Though the concept is now regarded as both basic and obvious, like the principle of competitive exclusion, it roused extraordinary opposition. The resistance to publication was the more poignant in that the young author was dying of an obscure hepatitis as the paper was finally accepted and went to press. The whole history has recently been recounted by Robert Cook. Thinking about the matter, and about a similar difficulty that befell my first graduate student, Gordon Riley, when he submitted a paper on plankton productivity, containing a great deal of statistical theory, to *Ecological Monographs,* I began to wonder whether he and Ray and I had not been suffering from a sort of common sense backlash generated at the Reformation by the ultra-intellectual and antiempirical aspects of medieval scholasticism, which backlash had flourished in America wherever Puritan attitude was still strong. I then remembered how when Prof. E. A. Birge and Professor Chancey Juday were kind enough to let me spend a week at the Trout Lake Laboratory in Vilas County, in northeastern Wisconsin, I had learned a fabulous amount about limnological technique but had come away with two feelings of dissatisfaction. One was that it would be nice to know how to put all their mass of data into some sort of informative scheme of general significance; the other was that it would be nice to have either tea or coffee, without seeming decadent and abnormal, for breakfast. I now suspect a connection."

18 Small springs were later studied by John Teal (1957) and by Laurence Tilly (1968). A large spring, Florida's Silver Spring, was studied by a team led by H. T. Odum, 1957.

19 Hutchinson (1979, 233) says that he came to biogeochemistry through Vernadsky. Hutchinson was introduced to George Vernadsky (V. I. Vernadsky's son) by Alexander Petrunkevitch, a professor at Yale.

Chapter 4: Transformation and Development of the Concept

1 Lotka (1925) would probably be the most important person to link physics and biology, yet (as mentioned in chap. 2) Lotka's contribution was not noted until much later. Juday (1940) never mentions Lotka in his energy budget article. Edward Haskell (1940) was quoted by Lindeman as an inspiration for a mathematical treatment of the ecosystem, yet he dropped the reference in his trophic-dynamics article. Haskell anticipated Hutchinson's concept of the niche. He quotes Eddington as saying that the whole world would be expressed in geometrical terms. Haskell then proposed that "the only kind of geometry to express this is n-dimensional geometry, where n is equal to the number of dimensions fitting the data. This region is equal to a geometric 'hyperbody,' defined as habitat. The hyperbody is identical with the physical construct of a field."

2 I have used Henry W. Spiegel (1971), *Growth of Economic Thought,* as a source on the history of economics. Donald Worster, in his book *Natures Economy* (1977), presents the view that ecology is the extension of economics to the whole world: "The weight of current interest lies not in hand-to-hand combat for survival, but with integrated circuitry, geochemical cycling, energy transfer. As a modernized economic system, nature now becomes a corporate state, a chain of factories, an assembly line." This new ecology derives from Elton's economic organization of nature and the use of energy as a currency. Worster, however, stresses that the new ecology is not merely the internal working of a science but reflects the larger cultural influences at work. A progressive conservation ethic, a faith in management, a materialist, utilitarian outlook are all considered formative.

3 By *top-down*, we mean that the system is analyzed into its parts and their interactions in order to explain the observed system behavior.

4 I am indebted to Eugene P. Odum for this account of the origin of *Fundamentals of Ecology.*

5 To illustrate the contrast of Odum's thought to that prevailing in conventional ecology, it is instructive to compare *Fundamentals* with Lee R. Dice's *Natural Communities* (1952). As indicated in the text, Dice's book presented a classical view of community ecology. Such statements as, "Unfortunately, however, no one person can ever have adequate training in all the fields of science that must be included in the consideration of even a single ecosystem" (p. 22); "It is doubtful, however, if any single index of community productivity can be found" (p. 143); and "Accurate prediction of the annual productivity of a given species in a particular ecosystem from a knowledge alone of the food or energy available and the physical conditions of the environment, however, is rarely possible" (p. 145) reveal how out of touch Dice was with the currents around him. *Natural Communities* might have been more relevant ten years earlier. By 1952, however, statements such as those cited here clearly no longer reflected current thinking.

6 Letter E. P. Odum to L. E. Norland, 14 April 1945, University of Georgia Archives, box 13, 1939–68.

7 A biography of Howard W. Odum appears in *Folk, Region and Society,* Jocher et al. (1964), which is mainly an anthology of Odum's work. The quotation is from the biography written by the editors (p. xi, xii). Taylor, 1988, 224.

8 In E. P. Odum's bibliography no papers deal with an ecosystem topic until 1955, when he and Tom Odum copublished "Trophic Structure and Productivity of a Windward Coral Reef Community on Eniwetok Atoll." Eugene Odum was deeply interested in lipid metabolism in birds, especially in relation to migration (see, e.g., Odum and Perkinson, 1951; Odum and Connell, 1956; Odum and Major, 1956) and the use of radioactive isotopes in relation to productivity and metabolism (see Odum, Kuenzler, and Blunt, 1958; Odum and Bachman, 1960; Odum and Pontin, 1961).

9 I have asked many ecologists from European countries who participated in the postwar redevelopment of the subject, Why was there so little involvement in ecosystem studies at this time? Two answers were frequently given. First, resources to build teams and research centers were not available. Second, the classical concepts of ecology, such as those of plant sociology and community classification, prevented development in this new way. We might add that, following Worster's interpretation (cited in note 2 here), European ecologists may not have been sufficiently influenced by a cultural milieu of economic utilitarianism and progressivism.

10 Worster (1977) gives a somewhat different interpretation to these trends. He emphasized the economic and managerial elements within ecosystem studies that shaped and were shaped by the dominant cultural environment of the time.

11 Odum published the second edition of *Fundamentals* in 1959, this time in collaboration with H. T. Odum. The second edition was 546 pages; the book had grown by 162 pages. New sections were added, such as a discussion of radiation ecology, and also new discoveries and information were added to each part. Thus, Odum kept up with the rapidly expanding field that he had so strongly influenced by his first edition.

12 This technique was apparently first used by Sargent and Austin (1949) to measure production and respiration of the coral reef at Ronglap Atoll, Marshall Islands. Odum and Odum (1955) cite it in their article on coral metabolism at Eniwetok.

13 An examination of the literature cited by H. T. Odum in articles published in the 1950s and early to mid-1960s reveals only a single reference to Lotka (1925) and none to Thienemann. Odum was referencing other literature in the German language, so Thienemann's ideas would have been accessible to him.

14 Eugene Odum (1953) was very clear about this matter: "Any entity or natural unit that includes living and nonliving parts interacting to produce a stable system in which the exchange of materials between living and nonliving parts follow circular paths is an ecological system or ecosystem" (p. 9); and, "The concept of ecosystem is and should be a broad one, its main function in ecological thought being to emphasize obligatory relationships, interdependence and causal relationships. Ecosystems may be conceived and studied in various sizes. Thus, the entire biosphere may be one vast ecosystem" (p. 10). Lamont Cole (1958) coined the word *ecosphere* for the planetary ecosystem.

15 Stanley Auerbach (1965), reviewing research in radioecology after about ten years of work began by stating, "Radionuclide cycling is an ecosystem process." His second paragraph starts with a reference to Lindeman. The fundamental ideas from ecosystem studies that could be used in radioecology were the food web and food chain, turnover rates, the circulation of biogeochemicals, and productivity. Auerbach's view was that the conceptual framework of modern ecosystem ecology was established in the early 1950s and applied in the mid-1950s to radiation problems. Thus, environmental research in the AEC, which began as a systematic documentation of the distribution of radioactivity in the environment, gradually shifted to a study of radioactive contamination in an ecological context.

16 The advantage of the Wiegert and Owens model is that it unpacks the decomposer component of the ecosystem and illustrates potential interactions between components.

17 I published an article (Golley, 1966) showing the relation between urban population density and the numbers of advertisements in the telephone books of Georgia cities. The advertisements were used to represent occupations. The relation was similar to that for biological species.

18 Lindeman had cited Haskell (1940) as his inspiration for using energy as the currency to express organism relationships in Cedar Bog Lake. In the trophic-dynamic article, the citation of Haskell is dropped and Thienemann (1918, 1926) is cited instead. Obviously, this is the more original citation. Lindeman

went beyond the use of energy as currency when he compared ratios of energy stored in or transferred between system components. The "efficiencies" could be used to compare the performance of different systems.

19 This chapter was written with the collaboration of his brother, H. T. Odum, and this assistance was specifically acknowledged in the preface of the second edition. The quotation is on page 65.

20 This is an example of a familiar process in the growth of scientific knowledge, as described by Diana Crane (1972). Precision of language is not the sole criterion for its use. Words such as *producer, consumer,* and *decomposer* had long antecedent usage and fit the common language, and therefore were easy to remember and use. Also, as Crane points out, ideas that are expressed by opinion leaders in a field, early in that field's development, have a stronger influence than newer ideas that may be more accurate or relevant. For example, Wiegert and Owen's revision of the Lindeman ecosystem model was made almost thirty years after Lindeman's publication of the tropic level concept, and while it was sound, it probably had less impact than it would have had, had it been made earlier.

21 Ecologists later developed the thermodynamic foundations for ecology. Two key articles were by D. Scott (1965) and Richard Wiegert (1968). The physicist Ichiro Aoki (1988) exactly defined the energy budget equation of Wiegert on the basis of the first law of thermodynamics.

22 Eugene Odum's style in the development of *Fundamentals* was to cite other authors to support or clarify a principle. He seldom cited the origin of an idea when a principle was presented or explained. While this style simplifies the text and makes it easier for a beginning student to understand the principles of ecology, it also leaves the reader with the impression that Odum derived the principles himself. Ordinarily, in a beginning text this style would cause no problem. However, *Fundamentals* was a new departure for ecology; it was the introduction to ecology for many scientists and administrators. It is unfortunate for our purposes that Odum chose this style, because it means that we cannot easily trace his sources.

23 The figure showing these data was published three times (Slobodkin and Richman, 1961; Slobodkin, 1961, 1962). These data were all obtained by Slobodkin and his students in his laboratory, using a microbomb calorimeter provided by the U.S. Navy. The bomb calorimeter that was in common use required about one-half gram of material for combustion. The microbomb calorimeter allowed much smaller samples to be analyzed.

The data I used for my compilation were obtained from a number of workers and represented plant and animal tissues, while all of Slobodkin and Richman's data were from animals. When my data are examined as a frequency distribution, they show a skewed pattern for the plants, with a mode of about 4,000 calories per gram dry weight, with values as high as 7,117 calories. The animal values do not show a skewed distribution and range from about 1,700 to 6,200 calories.

24 In 1946, a symposium on production in aquatic systems was published in *Ecologi-*

cal Monographs (16[4]). The articles on the dynamics of marine production by George Clarke and on the production of fish populations by William Ricker were exceptionally important and are cited by many later authors. Macfadyen cites Ricker's article as the source that attracted him to the problem of productivity.

25 As far as I know, Macfadyen is the first to emphasize that energy flow is unidirectional. The assertion that energy flows and materials cycle became a central tenet of ecology.

26 There were many attempts to compute the productivity of the entire earth. Gordon Riley's (1944) estimate was $146 \pm 87 \times 10^9$ tons, for an energetic equivalency of $13.6 \pm 8.1 \times 10^{20}$ calories. He calculated the mean efficiency of conversion of solar energy to carbon energy as 0.09 percent. Riley quotes several earlier estimates by German scientists.

27 The Savannah River Ecology Laboratory evolved from E. P. Odum's research project. In 1962, the laboratory was established and a resident staff hired. Before 1962, research was carried out mainly by graduate students working out of the University of Georgia, Athens. An exception was in the period 1956–58 when Robert Norris was the resident Ph.D. ecologist at the Savannah River Plant under the Odum program.

28 This idea repeats that of Tansley (1922, 25) in which he states that "all living organisms may be regarded as machines transforming energy from one form into another."

29 The number of ecologists in the Ecological Society of America during this period was approximately two thousand (Burgess, 1977).

Chapter 5: The International Biological Program

1 A biome is a biotic community of plants and animals, considered at a large geographic scale. The word *biome* is synonymous with formation or bioregion. According to Carpenter (1939), Clements proposed the term at the New York meeting of the Ecological Society of America as a synonym for biotic community. In 1932 Shelford applied it to large-scale communities.

2 In 1962, at the meeting of the International Association of Theoretical and Applied Limnology, in Madison, Wisconsin, W. Rodhe, who was the convener of the freshwater section of the IBP, presented a plan for freshwater studies. The delegates approved his plan. In 1965, at the next meeting in Warsaw, Poland, V. Tonolli asked for continued support of the IBP.

3 According to Rhodes W. Fairbridge (1966) the most recent era of marine exploration was termed the "era of international research cooperation." It began with the IGY (1957–59), and included the International Cooperative Investigations of the Tropical Atlantic and the International Indian Ocean Expedition.

4 Like many IBP meetings, this one was an exciting experience for the participants. For many it was their first opportunity to meet scientists from East to West, across cold-war boundaries. Kazimierz Petrusewicz, the chief of the Polish Academy of Science Institute of Ecology, was the host. The problems of defining the principles of production ecology required ad hoc night discussions, which attracted the serious attention of many participants. The fact that these discussions led to an eventual agreement added to the conviviality of the meeting. This spirit also made the selection of themes and choice of research topics easier. François Bourliere chaired the final formal meeting and by a process of voting with a show of hands, settled the difficult questions. Given the wide differences of opinion among botanists, zoologists, and scientists from different national groups, it was largely through Bourliere's leadership that a consensus was reached quickly.

5 At the Jablonna meeting, the interest of the scientists was largely in organisms, not habitats. The habitats provided a basis for comparison of the biota; it did not form the strongest element in the program. This orientation was continued through the IBP as studies of small mammals, granivorous birds, social insects, and so forth.

6 Vegetation was treated as the animal's food and habitat by the animal ecologists at Jablonna. Duvigneaud pointed out that vegetation can be subdivided vertically into strata, horizontally into life forms, and conceptually into species groups. A botanist could not be satisfied with PT committee's treatment of vegetation as equivalent to small mammals and social insects.

7 The survey of the U.S. scientific community by the NAS was carried out by an ad hoc committee chaired by Stanley Cain. American opinion was sampled by telephone and correspondence. One hundred and sixty-two scientists were consulted directly, and their names represent many of the most prominent and active ecologists, including L. C. Bliss, Francis Evans, Stanley Gessel, Donald Lawrence, Eugene Odum, H. T. Odum, Jerry Olson, William Osburn, Frank Pitelka, Arnold Schultz, L. B. Slobodkin, Robert Whittaker, and George Woodwell. In addition, the Ecological Society of America published a questionnaire in its bulletin (1963, 44:146–9) and had received 119 returns by February 1964. This questionnaire elicited suggestions for projects. Further, forty-one professional societies associated with the NAS Division of Biology and Agriculture were asked their opinion. Of the total respondents to the NAS telephone survey (162), 98 answered yes to the question, Should the United States participate in the IBP? 50 answered yes, if; and 7 answered no. The committee commented that the reluctance and opposition to an IBP arose from (1) the fear that the United States might be called upon to pay for the foreign programs, (2) the need for new money to be found, otherwise the ongoing programs would suffer financially, (3) that American research freedom might be placed in an international straitjacket, and (4) that efforts toward standardization might stifle originality (NAS/NRC 1944).

8 Proposal in the University of Georgia Archives, Odum, box 1.

9 In 1967 Roger Revelle was replaced by Frank Blair, a professor of zoology at the University of Texas.

10 Eugene Odum attributes the suggestion to Hasler. Neither Hasler nor Bliss recalls the source of the idea, but Hasler emphasized the concept "that lakes are

mirror images of the landscape around them" (Beckel, 1987, 31), and he suggested that he might well have expressed this view when the organization was in a formative stage (Halser, letter to Golley, 21 Dec. 1988).

11 University of Georgia Archives, Odum, box 1.

12 Ibid.

13 Proposal for a PT and PF program, 31 Sept. 1966. University of Georgia Archives, Odum, box 1.

14 Ibid., box 13.

15 Progress Report, Analysis of Ecosystems, 27 June 1967 by Fred Smith. University of Georgia Archives, Odum, box 13.

16 Letter from Philip Johnson to the faculty at the University of Georgia Institute of Ecology, 8 Jan. 1968. University of Georgia Archives, Odum, box 13.

17 First Annual Report "Analysis of Ecosystems," 11 Nov. 1967. University of Georgia Archives, Odum, box 13.

18 Proposal to the NSF by Colorado State University, Dec. 1967, pp. E1–22. University of Georgia Archives, U.S. IBP Grassland Biome, Budget 1968–69, proposal 1968–69, Progress Report, Continuation Proposal 1969–70.

19 University of Georgia Archives, U.S. IBP Grassland Biome, Budget 1968–69, proposal 1968–69, Progress Report, Continuation Proposal 1969–70.

20 Proposal for Continuation of research grant GB-7824 from Colorado State University, 135–156. U.S. IBP Grassland Biome, Budget 1968–69, proposal 1968–69, Progress Report, Continuation Proposal 1969–70. University of Georgia Archives.

21 For example, the premature review of the IBP biome studies by Mitchell, Mayer, and Downhower (1976) contrasted the productivity of publications from the Hubbard Brook project with those from three IBP projects.

22 Proposal for Continuation of research grant GB-31862X and GB-31862X2 from Colorado State University, June 1973. U.S. IBP Grassland Biome Continuation Proposal 1974–76, Progress Report 1973, Annual Report 1974–75. University of Georgia Archives.

23 Proposal to the NSF by Colorado State University, Dec. 1967. University of Georgia Archives. U.S. IBP Grassland Biome, Budget 1968–69, proposal 1968–69, Progress Report, Continuation proposal 1969–70.

24 Proposal for Continuation of Research grant GB-7824 from Colorado State University. U.S. IBP Grassland Biome, Budget 1968–69, proposal 1968–69, Progress Report, Continuation Proposal 1969–70. University of Georgia Archives.

Chapter 6: Consolidation and Extension of the Concept

1 The background about the development of the Hubbard Brook project comes from remembrances of Herbert Bormann printed in 1985, and Likens, 1985.

2 Likens et al., 1977. The 1975 review of the three biomes (Battelle Columbus, 1975) lists 107 articles published by the Hubbard Brook team.

3 Introduction to the Ellenberg *Festschrift* by Wolfgang Haber, 9–13, in Schmidt, 1983.

4 For example, see Kato, Tadaki, and Ogawa, 1978, in a special issue of the *Malayasian Nature Journal* for a report on plant productivity at Pasoh.

5 See Colwell (1973), in an unpublished technical report available in the University of Georgia Archives, IBP Collection, *Origin and Structure of Ecosystems*.

6 See the review by Cody and Mooney (1978), for a general discussion of convergence.

7 The society is called the Mediterranean Ecological Society (MEDECO) and meets at approximately two-year intervals.

8 Mar, 1977. This was a review funded by NSF for purposes of evaluating the RES program.

9 Di Castri returned to UNESCO in 1990.

10 Herrera et al. (1978) presents a useful summary and a list of relevant supporting articles to the San Carlos project.

Chapter 7: Interpretations and Conclusions

1 John Algeo (1988) describes numbers of *eco-* words that have become current in American literature and speech. These words include *ecotage, ecofix, ecopornography, ecodisaster, ecodefense, ecofact, ecomenu, econote,* and so on. The usage almost always implies some connection with the environment. Thus, *ecodisaster* would mean a serious event in which a disaster occurred in the environment, causing environmental disturbance. The use of *eco-* for environment comes from misuse of ecology as a synonym for environment by the American media beginning in the late 1960s. Apparently, the shorter word *ecology* fit the column width of a printed page better than the longer word *environment*. Newspapers especially were impervious to repeated attempts by ecologists to correct this misusage. It is now fixed in the language.

2 The list is that of Sukachev and Dylis (1964), 16, which is cited in their discussion over the advantages of using biogeocenosis in place of ecosystem and facies.

3 For example, the human body was referred to as "that magnificent machine" in a popular television film. Houses are termed "machines for living" in architectural writing. This conception may have originated in the form of a clocklike mechanical world, stimulated by the discoveries of Galileo, Descartes, and Newton, among others, and continues today in the form of a computer or space colony (Ferré, 1988). Although in the environmentalist literature there is strong condemnation for using a machine metaphor for living objects or ecosystems, nevertheless the metaphor has wide popular appeal.

4 Weiner, 1988, 132. See also Gershenson (1990, 450): "Lysenko's chief assistant

and supporter, I. I. Prezent, was a lawyer. Lysenko recommended him for a professorship of biology and he simultaneously held chairs in both the Moscow and
Leningrad Universities."

5 Although I have no direct evidence of a connection of Thienemann with national
socialism, Soderquist (1986) implies such a connection (p. 269). Trommer,
1990, and in conversation, pointed out that national socialism was, in a sense,
applied biology. Ecologists worked with Nazi developers of large industry, the
autobaum, and other developments to harmonize them. Biological sciences had
been suppressed in the late 1800s in the German high schools because of the opposition of the Church to Darwinism. Under the Nazis this situation was
changed, biology was valued and became an important subject in schools supported by the Nazi party.

6 David Gates lectured widely during the period when he served as director of the
Missouri Botanical Garden, after he published his small book on energy exchange
in the biosphere (Gates, 1962). During a lecture at Emory University, Gates
forcefully criticized ecology as being too focused on biological factors and ignoring the physical-chemical relations in the environment.

7 Geography, defined as regional science, was supported in one program of NSF,
with a budget of about $3 million annually in 1991.

8 Numata (1990) describes the development of ecology in Japan. My discussion is
based largely on Numata and on conversations with Numata.

9 An example is Fenchels's (1987) comment: "Another, it seems now extinct, approach was to make analogies between nonequilibrium thermodynamics and
ecological systems, such as equating species diversity of communities with 'negative entropy' of chemical systems. Again, this is a fundamentally false analogy and
it had an appeal for some time, I suppose, only because it was sufficiently obscure
and incomprehensible to appear profound" (p. 17).

10 Holism, as mentioned in chap. 2, was developed by the South African statesman
and general Jan Christian Smuts (1926) in a book, titled *Holism and Evolution*.
Smuts's concept of holism was based on a process of evolution of the physical
molecule to the individual and ultimately to mind and value. The approach has
similarity to that of Pierre Teilhard de Chardin in *The Phenomenon of Man*. Smuts
emphasized that the individual was a whole and that holism was the force that
creates wholes. This concept, which is considerably different from the holism
concept used in modern ecology, came into ecology through John Phillips. Actually, Smuts never speaks of ecological communities in his book. The nearest he
comes are several statements: "Taking all the wholes in the world and viewing
them together in Nature, we see a similar interpenetration and enrichment of the
common field. When we speak of nature we do not mean a collection of uncon-
nected items, we mean wholes with their interlocking fields; we mean a creative
situation which is far more than the mere gathering of individuals and their separate fields. . . . But the sober fact is that there is no new whole or organism of
Nature; there is only Nature become organic through the intensification of her

total field. In other words, Nature is holistic without being a real whole" (Smuts, 340). And: "The holistic organic field of Nature exercises a similar subtle moulding, controlling influence in respect of the general trend of organic advance. That trend is not random or accidental or free to move in all directions; it is controlled, it has the general character of uniform direction under the influence of the organic or holistic field of Nature" (p. 342). Phillips seems to twist Smuts's meaning in his reference (Phillips, 1931, 20): "While I—and doubtless Clements himself—would agree that philosophically General Smuts (339–43) by his masterly and inspiring exposition—in a universal connection—that groups, societies, nations, and Nature are *organic without being organisms,* are holistic without being wholes—has pointed to the truth, I still am able to see that the concept of the *complex organism* has much to commend it in practice. . . . In accordance with the holistic concept of Smuts, the biotic community is something more than the mere sum of its parts; it possesses a special identity—it is indeed a mass-entity with a destiny peculiar to itself." In this conclusion Phillips is defending the superorganism concept.

REFERENCES

Adams, Charles C. 1915. An outline of the relations of animals to their inland environments. *Bull. Illinois State Laboratory Natural History* 11(1): 1–32.

———. 1918. Migration as a factor in evolution: Its ecological dynamics. *Am. Naturalist* 52(622, 623): 465–90.

———. 1919. Migration as a factor in evolution: Its ecological dynamics II. *Am. Naturalist* 53:55–78.

Algeo, John. 1988. Among the new words. *American Speech* 63:345–52.

Allee, W. C., Alfred E. Emerson, Orlando Park, Thomas Park, and Karl P. Schmidt. 1949. *Principles of Animal Ecology*. W. B. Saunders, Philadelphia.

Allen, T. F. H., and Thomas B. Starr. 1982. *Hierarchy: Perspectives for Ecological Complexity*. Univ. of Chicago. Press, Chicago.

Alsterberg, G. 1922. Die respiratorischen Mechanismen der Tubificiden: Eine experimentell-physiologische Untersuchung auf ökologischer Grundlage. *Lunds Univer. Arsskrift* 11(18): 1–176.

———. 1925. Die Nahrungszirkulation einiger Binnenseetypen. *Arch. Hydrobiol.* 15:291–338.

Andreski, Stanislav. 1971. *Herbert Spencer: Structure, Function, and Evolution*. Michael Joseph, London.

Andrewartha, H. G., and L. C. Birch. 1954. *The Distribution and Abundance of Animals*. Univ. of Chicago Press, Chicago.

Aoki, Ichiro. 1988. Exact derivation of an energy budget equation on the basis of the first law of thermodynamics. Ecol. *Research* 3:53–56.

Auerbach, S. I. 1965. Radionuclide cycling: Current status and future needs. *Health Physics* 11:1355–61.

———. 1970. Analysis of the structure and function of ecosystems in the deciduous forest biome. Proposal to the NSF, 1 July 1970, Research Design, 14.

Baird, Lewis C. 1909. *Baird's History of Clarke County, Indiana*. B. F. Bowen, Indianapolis.

Baladin, R. K. 1982. *Vladimir Vernadsky: Outstanding Soviet Scientists*. Mir, Moscow.

Barnhardt, Robert K. 1988. *The Barnhardt Dictionary of Etymology*. H. G. Wilson.

Battelle Columbus Laboratories. 1975. Final Report to the NSF for evaluation of three of the biome studies programs funded under the foundation's international biological program (IBP).

Beckel, A. L. 1987. *Breaking New Water: A Century of Limnology at the University of Wisconsin*. Special Issue, Transactions of the Wisconsin Academy of Science, Arts and Letters.

Bergson, Henri. 1911. *Creative Evolution*. H. Holt, New York.

Bernard, Claude. 1878. *La science experimentale*. Paris.

Berry, R. J. 1988. Natural history in the twentieth-first century. *Archives of Natural History* 15:1–14.

Bertalanffy, Ludwig von. 1950. An outline of general system theory. *Brit. J. Philosophy Science* 1:134–65.

———. 1952. *Problems of Life: An Evaluation of Modern Biological and Scientific Thought*. Harper Torchbooks, New York.

Beyers, Robert J. 1963. The metabolism of twelve aquatic laboratory micro-ecosystems. *Ecol. Monographs* 33:281–306.

Birch, L. C., and D. P. Clark. 1953. Forest soil as an ecological community, with special reference to the fauna. *Quar. Rev. Biol.* 28:13–36.

Birge, E. A. 1907. The respiration of an inland lake. *Popular Science Monthly* 72:337–51. (Also in *Trans. Am. Fish. Soc.* 36:223–41.)

———. 1910. On the evidence for temperature seiches. *Trans. Wisconsin Academy Science, Arts and Letters* 16:1005–16.

Birge, E. A., and C. Juday. 1911. The inland lakes of Wisconsin: The dissolved gases and their biological significance. *Bull. Wisc. Geol. Nat. Hist. Surv.* 22.

———. 1934. Particulate and dissolved organic matter in inland lakes. *Ecol. Monographs* 4:440–74.

Blackman, F. F., and A. G. Tansley. 1905. Ecology in its physiological and phytotopographical aspects. *New Phytologist* 4(9): 199–203, 232–53.

Blair, W. Frank. 1977. *Big Biology*. The US/IBP. Dowden, Hutchinson and Ross, Stroudsburg.

Bohm, David. 1980. *Wholeness and the Implicate Order*. Routledge & Kegan Paul, London.

Bonner, J. 1962. The upper limit of crop yield. *Science* 137:11–15.

Bormann, F. H. 1985. *Lessons from Hubbard Brook*. Report 62, California Water Resources Center, Univ. California, Davis.

Bormann, F. H., and G. E. Likens. 1967. Nutrient cycling. *Science* 155(3461): 424–29.

———. 1979. *Pattern and Process in a Forested Ecosystem*. Springer-Verlag, New York.

Bormann, F. H., G. E. Likens, T. G. Siccama, R. S. Pierce, and J. S. Eaton. 1974. The export of nutrients and recovery of stable conditions following deforestation at Hubbard Brook. *Ecol. Monographs* 44(3): 255–77.

Borutsky, E. V. 1939. *Dynamics of the Total Benthic Biomass in the Profundal of Lake Beloie*. Proc. Kossino Limnol. Station, Hydrometeorological Service USSR 22:196–218. Trans. M. Ovchynnyk, ed. C. Ball, F. F. Hooper, Mich. State. Univ.

Bramwell, Anna. 1985. *Blood and Soil: Richard Walther Darré and Hitler's 'Green Party.'* The Kensal Press. Abbotsbrook, Bourne End, Buckinghamshire.

———. 1989. *Ecology in the Twentieth Century*. Yale Univ. Press, New Haven.

Braun-Blanquet, J. 1932. *Plant Sociology: The Study of Plant Communities*. McGraw-Hill, New York. (Trans. G. D. Fuller and H. S. Conrad.)

Bray, J. R. 1963. Root production and the estimation of net productivity. *Canadian J. Bot.* 41:65–72.

Bray, J. R., D. B. Lawrence, and L. C. Pearson. 1959. Primary production in some Minnesota terrestrial communities for 1957. *Oikos* 10:38–49.

Brennan, Andrew. 1988. *Thinking about Nature: An Investigation of Nature, Value and Ecology*. Univ of Georgia Press, Athens.

Breymeyer, A. I., and G. M. Van Dyne (eds.). 1980. *Grasslands, Systems Analysis and Man*. Cambridge Univ. Press, Cambridge.

Briggs, John, and F. David Peat. 1984. *Looking Glass Universe: The Emerging Science of Wholeness*. Simon & Schuster, New York.

Brock, Thomas D. 1967. The ecosystem and the steady state. *Bioscience* 17:166–69.

Brown, J., and H. N. Coulombe. 1969. Analysis of the structure and function of the wet tundra ecosystem at Barrow, Alaska. U.S. IBP Biome Barrow Workshop, Univ. Colorado, 13.

Bunge, Mario. 1979. *Treatise on Basic Philosophy*. Vol. 4. Ontology II, A World of Systems. D. Reidel, Dordrecht.

Burgess, Robert L. 1977. The Ecological Society of America, Historical Data and some preliminary analyses. In F. N. Egerton and R. P. McIntosh (eds.), *History of American Ecology*. Arno Press, New York.

Bush, Vannevar. 1945. *Science: The Endless Frontier*. United States Office of Scientific Research and Development, Washington, D.C.

Cahn, Harley. 1988. Against the moral considerability of ecosystems. *Environmental Ethics* 10(3): 195–216.

Cale, W. G. 1975. *Simulation and Systems Analysis of a Shortgrass Prairie Ecosystem*. Ph.D. diss., Univ. of Georgia, Athens.

Callicott, J. Baird. 1990. The case against moral pluralism. *Environmental Ethics* 12(2): 99–124.

Cannon, Walter. 1932. *The Wisdom of the Body*. New York.

Carney, H. J. 1989. On competition and the integration of population, community and ecosystem studies. *Function Ecology* 3:637–41.

Carpenter, J. R. 1939. The biome. *Am. Midland Naturalist* 21:75–91.

Carr, William. 1969. *A History of Germany, 1815–1945*. St. Martins, New York.

Carson, Rachel. 1962. *Silent Spring*. Houghton Mifflin, Boston.

Cherrett, J. M. 1989. Key concepts: The results of a survey of our members' opinions, pp. 1–16. In J. M. Cherrett (ed.), *Ecological Concepts: The Contribution of Ecology to an Understanding of the Natural World*. Blackwell, Oxford.

Clarke, F. W. 1908. *Data of Geochemistry*. Bull. U.S. Geological Survey. No. 770.

Clarke, George L. 1946. Dynamics of production in a marine area. *Ecol. Monographs* 16:322–35.

———. 1954. *Elements of Ecology*. Wiley, New York.

Clarke, G. L., W. T. Edmondson, and W. E. Ricker. 1946. Mathematical formulation of biological productivity. *Ecol. Monographs* 16:336–37 (app.).

Clements, Frederic E. 1905. *Research Methods in Ecology*. Lincoln, Nebraska.

———. 1916. *Plant Succession: An Analysis of the Development of Vegetation*. Carnegie Institution, Washington.

Clements, Frederic E., C. T. Vorhies, and W. P. Taylor. 1922. *Plant Competition*. Carnegie Institution, Washington, D.C., Yearbook 21:355.

Clements, Frederic E., and Victor E. Shelford. 1939. *Bio-Ecology*. Wiley, New York.

Cloud, Preston. 1974. Evolution of ecosystems. *Am. Scientist* 62:54–66.

Cody, M. L., and H. A. Mooney. 1978. Convergence versus nonconvergence in Mediterranean-climate ecosystems. *Ann Rev. Ecology and Systematics* 9:265–321.

Cole, Lamont C. 1958. The Ecosphere. *Scientific American* (April): 1–7.

Colwell, Robert. 1973. *Origin and Structure of Ecosystems.* Technical Report 73–78. IBP Origin and Structure of Ecosystems Berkeley Workshop, Univ. of California, Berkeley, 21–23 Jan. 1973.

Connell, Joseph H., and Eduardo Orias. 1964. The ecological regulation of species diversity. *Am. Naturalist* 98:399–414.

Cook, Robert Edward. 1977. Raymond Lindeman and the trophic-dynamic concept in ecology. *Science* 198:22–26.

Cooper, W. S. 1923. The recent ecological history of Glacier Bay, Alaska II: The present vegetation cycle. *Ecology* 4:223–46.

Cowles, H. C. 1899. The ecological relationships of the vegetation on the sand dunes of Lake Michigan. *Botanical Gazette* 27:95–391.

———. 1901. The Physiographic Ecology of Chicago and Vicinity: A study of the origin, development, and classification of plant societies. *Botanical Gazette* 31:73–108, 145–82.

Crane, D. 1972. *Invisible Colleges: Diffusion of Knowledge in Scientific Communities.* Univ. of Chicago Press, Chicago.

Croker, Robert A. 1991. *Pioneer Ecologist: The Life and Work of Victor Ernest Shelford, 1877–1968.* Smithsonian Institution Press, Washington, D. C.

Cummings, K. W., and J. C. Wuycheck. 1971. Caloric equivalents for investigations in ecological energetics. *Mitteilungen Internationale Vereinigung Limnologie* 18.

Darnell, Rezneat M. 1970. Evolution and the ecosystem. *Am. Zoologist* 10:9–15.

Davidson, Mark. 1983. *Uncommon Sense: The Life and Thought of Ludwig von Bertalanffy (1901–1972), Father of General Systems Theory.* J. B. Tarcher, Los Angeles.

Davis, M. B. 1981. Quaternary history and the stability of forest communities, pp. 132–53. In D. C. West, H. H. Shuggart, and D. B. Botkin (eds.), *Forest Succession, Concepts and Application.* Springer-Verlag, New York.

Degrood, David H. 1965. *Haeckel's Theory of the Unity of Nature.* Christopher, Boston.

Delcourt, H. R., and P. A. Delcourt. 1988. Quaternary landscape ecology: Relevant scales in space and time. *Landscape Ecology* 2(1): 23–44.

Detling, J. K., M. I. Dyer, C. Procter-Gregg, and D. T. Winn. 1980. Plant-herbivore interactions, examination of potential effects of buffalo saliva on regrowth of *Bouteloua gracilis* (H. B. K.) *Lag. Oecologia* 45:26–31.

Di Castri, F. 1983. L'écologie. *Les défis d'une science en temps de crise. Ministre de l'Industrie et de la Recherche.* La Documentation Française.

Dice, Lee R. 1952. Natural Communities. Michigan Univ. Press, Ann Arbor.

Dineen, Clarence F. 1953. An ecological study of a Minnesota pond. *Am. Midland Naturalist* 50:349–76.

Drude, Oscar. 1895. *Deutschlands Pflanzengeographie.* Braunschweig.

Dyer, Melvin I. 1980. Mammalian epidermal growth factor promotes plant growth. *Proc. Nat. Acad. Sci.* 77(8): 4836–37.

Egerton, Frank N. 1973. Changing concepts of the balance of nature. *Quarterly Review of Biology* 48:322–50.

Ehrlich, P. R., and L. C. Birch. 1967. The "balance of nature" and "population control." *Am. Naturalist* 101:97–107.

Ellenberg, Heinz. 1971. *Integrated Experimental Ecology: Methods and Results of Ecosystem Research in the German Solling Project.* Springer-Verlag, Berlin.

Ellenberg, H., R. Mayer, and J. Schauermann. 1986. *Ökosystemforschung Ergebnisse des Sollingprojekts, 1966–1986.* Eugen Ulmer, Stuttgart.

Elster, Hans-Joachim. 1974. History of limnology. *Mitt. Int. Verein. Limnol.* 20:7–30.

Elton, Charles. 1927. *Animal Ecology.* Sidgwick and Jackson, London. Reprinted in 1939, Macmillan, New York.

———. 1966. *The Pattern of Animal Communities.* Wiley, New York.

Engelberg, J., and L. L. Boyarsky. 1979. The noncybernetic nature of ecosystems. *Am. Naturalist* 114:317–24.

Engelmann, Manfred D. 1961. The role of soil arthropods in the energetics of an old field community. *Ecol. Monographs* 31:221–38.

Evans, Francis C. 1956. Ecosystem as the basic unit in ecology. *Science* 123:1127–28.

Evans, F. C., and E. Dahl. 1955. The vegetation structure of an abandoned field in southeastern Michigan and its relation to environmental factors. *Ecology* 36:685–706.

Evans, G. Clifford. 1976. A sack of uncut diamonds: The study of ecosystems and the future resources of mankind. *J. Ecology* 64:1–39.

Fairbridge, Rhodes W. (ed.) 1966. *The Encyclopedia of Oceanography.* Reinhold, New York.

Fenchel, Tom. 1987. *Ecology: Potentials and Limitations.* Excellence in Ecology 1. Ecology Institute, Oldendorf/Luhe.

———. 1989. Comment on Carney's article by T. Fenchel. *Functional Ecology* 3:641.

Fenton, G. R. 1947. The soil fauna, with special reference to the ecosystem of forest soil. *J. Anim. Ecol.* 16:76–93.

Ferré, Frederick. 1988. *Philosophy of Technology*. Prentice Hall, Englewood Cliffs.

Fisher, R. A., A. S. Corbett, and C. B. Williams. 1943. The relation between the number of species and the number of individuals in a random sample of an animal population. *J. Anim. Ecol.* 12:42–58.

Folk, E. G., Jr., and R. S. Hedge. 1964. Comparative physiology of heart rate of unrestrained mammals. *Am. Zool.* 4:111.

Fontaine, Y. 1960. *Radioactive Contamination of Aquatic Media and Organisms.* AEC Translation 5358 of Report CEA-1588, Comm. à l'Energie Atomique. Centre d'Etudes Nucléaires de Salcay, France.

Forbes, S. A. 1887. The lake as a microcosm. Bull. Peoria Sci. Ass., *Illinois Nat. Hist. Surv. Bull.* 15:537–50.

———. 1907. History of the former state Natural History Societies of Illinois. *Science* 26(678): 892–98.

Forel, F. A. 1892, 1895, 1907. *Le Leman: Monographie limnologique.* 3 vols. Lausanne, F. Rouge.

Fortescue, John. 1992. Landscape geochemistry: Retrospect and prospect, 1990. *Applied Geochemistry* 7:1–53.

French, N. (ed.). 1979. *Perspectives in Grassland Ecology, Results and Applications of the U.S. IBP Grassland Biome Study.* Springer-Verlag, New York.

French, N., R. K. Steinhorst, and D. M. Swift. 1979. Grassland biomass trophic pyramids, pp. 59–87. In N. French (ed.), *Perspectives in Grassland Ecology.* Springer-Verlag, New York.

Frey, David G. 1963. Wisconsin: The Birge-Juday era, pp. 3–54. In David G. Frey (ed.), *Limnology in North America.* Univ. Wisconsin Press, Madison.

Frey, T. 1977. *Spruce Forest Ecosystem Structure and Ecology.* 1. Introductory Data on the Estonian Vooremaa Project. Acad. Sci. Estonian SSR, Tartu.

———. 1979. *Spruce Forest Ecosystem Structure and Function.* 2. Basic Data on the Estonian Vooremaa Project. Acad. Sci. Estonian SSR, Tartu.

———. 1981. *Structure and Ecology of Temperate Forest Ecosystems.* Acad. Sci. Estonian SSR, Tartu.

Friederichs, K. 1927. Grundsätzliches über die Lebenseinheiten höherer Ordnung und den ökologischen Einheitsfaktor. *Die Naturwissenschaften* 15:153–57, 182–286.

———. 1937. Ökologie als Wissenschaft von der Natur oder biologische Raumforschung. *Bios.* 7. Leipzig.

232

Gasman, Daniel. 1971. *The Scientific Origins of National Socialism: Social Darwinism in Ernst Haeckel and the German Monist League*. MacDonald, London; American Elsevier, New York.

Gates, David M. 1962. *Energy Exchange in the Biosphere*. Harper and Row, New York.

Gershenson, S. M. 1990. The grim heritage of Lysenkoism. Four personal accounts. 4, Difficult years in Soviet Genetics. *Q. Rev. Biol.* 65(4): 447–56.

Gibbs, J. W. 1878. On the equilibrium of heterogeneous substances. *Connecticut Academy Transactions* 3:108–248, 343–524; *Am. J. Science* 16:441–58.

Glass, Bentley. 1964. The critical state of the critical review article. *Q. Rev. Biol.* 39:182–85.

Gleason, H. A. 1917. The structure and development of the plant association. *Bull. Torrey Botanical Club* 43:463–81.

———. 1922. On the relation between species and area. *Ecology* 3:158–62.

———. 1926a. The individualistic concept of the plant association. *Bull. Torrey Botanical Club* 53:7–26.

———. 1926b. *Plant Associations and Their Classification: A Reply to Dr. Nichols*. Proc. Int. Congress of Plant Science, Ithaca.

———. 1939. The individualistic concept of the plant association. *Am. Midland Naturalist* 21:92–110.

Godwin. 1957. *Arthur George Tansley, 1871–1955*. Biographical Memoirs, Royal Society. London.

———. 1977. *Sir Arthur Tansley: The Man and the Subject. J. Ecology* 65:1–26.

———. 1985. *Cambridge and Clare*. Cambridge Univ. Press, Cambridge.

Publications by Frank B. Golley

1960. Energy dynamics of an old-field community. *Ecol. Monographs* 30:187–206.

1961. Energy values of ecological materials. *Ecology* 42:581–84.

1962. The eight-year trend in quail and dove call counts in the AEC Savannah River Plant Area. *Trans. 27th N.A. Wildlife and Nat. Res. Conf.*, pp. 212–24. Wildlife Mgt. Inst., Washington, D.C.

1965. Structure and function of an old-field broomsedge community. *Ecol. Monographs* 35:113–31.

1966. The variety of occupations in human communities compared with the variety of species in natural communities. *Bull. Ga. Acad. Sci.* 24(1): 1–6.

1967. Methods of measuring secondary productivity in terrestrial vertebrate populations, pp. 99–124. In K. Petrusewicz (ed.), *Secondary Productivity of Terrestrial Ecosystems*. Inst. Ecol., PAN, Warsaw.

1972. Energy flux in ecosystems, pp. 69–90. In J. A. Wiens (ed.) *Ecosystem Structure and Function*. Oregon State Univ. Press, Corvallis.

1974. Structural and functional properties as they influence ecological stability, pp. 97–102. In *Proc. First Int. Congress of Ecology*. Centre for Agric. Publ. and Documentation, Wageningen.

1980. Report on the Division of Environmental Biology, NSF. *Bull. Ecological Society of America* 61(4): 210–14.

1984. Historical origins of the ecosystem concept in biology, pp. 33–50. In E. Moran, *The Ecosystem Concept in Anthropology*. Westview Press, Boulder.

1987. Deep ecology from the perspective of environmental science. *Environmental Ethics* 9(1): 45–55.

Grodzinski, Wladyslaw, and Andrzej Gorecki. 1967. Daily energy budgets of small rodents, pp. 295–314. In K. Petrusewicz (ed.), *Secondary Productivity of Terrestrial Ecosystems*. Polish Acad. Sci., Warsaw.

Grodzinski, Wladyslaw, and Bruce Wunder. 1975. Ecological energetics of small mammals, pp. 173–204. In F. B. Golley, K. Petrusewicz, and L. Ryszkowski (eds.), *Small Mammals: Their Productivity and Population Dynamics*. Cambridge Univ. Press, Cambridge.

Gustafson, Philip E. 1966. Environmental radiation. *Argonne National Laboratory Reviews* 3(3): 67–74.

Haeckel, Ernst. 1866. *Generelle Morphologie der Organismen: Allgemeine Grundzüge der organischen Formen-Wissenschaft, mechanisch begründet durch die von Charles Darwin reformierte Descendenz-Theorie*. 2 vols. Reimer, Berlin.

Hagen, Joel. 1992. *An Entangled Bank: The Origins of Ecosystem Ecology*. Rutgers Univ. Press, New Brunswick.

Hairston, Nelson G. 1959. Species abundance and community organization. *Ecology* 40:404–16.

Hairston, Nelson G., and George W. Byers. 1954. The soil arthropods of a field in southern Michigan: A study in community ecology. *Contr. Lab. Vert. Biol.* 64:1–37.

Hairston, Nelson G., Frederick E. Smith, and Lawrence B. Slobodkin. 1960. Community structure, population control, and competition. *Am. Naturalist* 94:421–25.

Hargrove, E. C. 1989. *Foundations of Environmental Ethics*. Prentice Hall, Englewood Cliffs.

Harper, J. L. 1977a. *The Population Biology of Plants*. Academic Press, New York.

――――. 1977b. The contributions of terrestrial plant studies to the development of the theory of ecology, pp. 139–57. In C. E. Goulden (ed.), *The Changing Scenes in the Natural Sciences: 1776–1976*. Acad. Nat. Sci., Philadelphia.

Hartmann, M. 1924. Allgemeine Biologie, 1st ed., G. Fischer, Stuttgart.

Hartmann, Nicolai. 1921. Grundzüge einer Metaphysik der Erkenntnis. DeGruyter, Berlin.

Haskell, Edward F. 1940. Mathematical systematization of "environment," "organism" and "habitat." *Ecology* 21(1): 1–16.

Hensen, V. 1887. Über die Bestimmung des Planktons oder des im Meere treibenden Materials an Pflanzen und Tieren. *Berlin Komm. wiss. Unters. dt. Meere* 5:1–109.

Herrera, Rafael, C. F. Jordan, H. Klinge, and E. Medina. 1978. Amazon ecosystems. Their structure and functioning with particular emphasis on nutrients. *Interciencia* 3(4): 223–32.

Hiratsuka, E. 1920. Researches on the nutrition of the silk worm. *Bull. Sericult. Exp. Sta. Japan*. 1:257–315.

Hollings, C. S. 1965. The functional response of predators to prey density and its role in mimicry and population regulation. *Mem. Ent. Soc. Canada* 45:3–60.

Hutchinson, G. Evelyn. 1964. The lacustrine microcosm reconsidered. *Am. Scientist* 52:334–41.

――――. 1979. *The Kindly Fruits of the Earth: The Development of an Embryo Ecologist*. Yale University Press, New Haven.

Innis, George S. 1978. Objectives and structures for a grassland simulation model, pp. 1–21. In Innis, *Grassland Simulation Model*. Springer-Verlag, New York.

Ivlev, V. S. 1945. The biological productivity of waters. *Uspekhi Sovremennoi Biologii* (Advances in Modern Biology) 19(1): 98–120. Trans. W. E. Ricker.

Ivorson, L. R. 1988. Land-use changes in Illinois, USA: The influence of landscape attributes on current and historic land use. *Landscape Ecology* 2:45–62.

Ivorson, L. R., R. L. Graham, and E. A. Cook. 1989. Applications of satellite remote sensing to forested ecosystems. *Landscape Ecology* 3(2): 131–43.

Iwaki, Hideo. 1958. The influence of density on the dry matter production of *Fagopyrum esculentum*. *Jap. J. Botany* 16:210–26.

Jocher, K., G. B. Johnson, G. L. Simpson, and R. B. Vance (eds.). 1964. *Folk, Region and Society: Selected Papers of Howard W. Odum*. Univ. North Carolina Press, Chapel Hill.

Jordan, Carl F. 1981. Do ecosystems exist? *Am. Naturalist* 118:284–87.

―――. 1987. *Amazonian Rain Forests: Ecosystem Disturbance and Recovery*. Springer-Verlag, New York.

Juday, Chancey. 1940. The annual budget of an inland lake. *Ecology* 21(4): 438–50.

Kato, R., Y. Tadaki, and H. Ogawa. 1978. Plant biomass and growth increment studies in Pasoh Forest. *Malay. Nat. J.* 30(2): 211–24.

Kerner, Anton von Marilaun. [1863] 1951. *The Plant Life of the Danube Basin*. Trans. H. S. Conrad and titled *The Background of Plant Ecology*. Iowa State Univ. Press, Ames.

―――. 1897. *Natural History of Plants: Their Forms, Growth, Reproduction and Distribution*. Trans. F. W. Oliver. Blackie, London.

Kimler, William C. 1986. Advantage, adaptiveness and evolutionary ecology. *J. Hist. Biol.* 19(2): 215–33.

Kingsland, Sharon E. 1985. *Modeling Nature*. Univ. of Chicago Press, Chicago.

Kira, T., and T. Shidei. 1967. Primary production and turnover of organic matter in different forest ecosystems of the western Pacific. *Jap. J. Ecol.* 17:30–87.

Kitazawa, Y. (ed.). 1977. *Ecosystem Analysis of Subalpine Coniferous Forest of the Shigayama* IBP Area, Central Japan. JIBP Synthesis. Vol. 15. Tokyo.

Koestler, Arthur. 1978. *Janus: A Summing Up*. Random House, New York.

Kozlovsky, Daniel G. 1968. A critical evaluation of the trophic level concept. I. Ecological efficiencies. *Ecology* 49:48–60.

Kuhn, Thomas S. 1962. *The Structure of Scientific Revolutions*. Univ. of Chicago Press, Chicago.

Kuroiwa, Sumio. 1960a. Ecological and physiological studies of the vegetation of Mt. Shimagare. IV. Some physiological functions concerning matter production in young *Abies* trees. *Botanical Magazine* (Tokyo) 73:133–41.

―――. 1960b. Ecological and physiological studies on the vegetation of Mt. Shimagare. V. Intraspecific competition and productivity difference among tree classes in the *Abies* stand. *Botanical Magazine* (Tokyo) 73:165–74.

Lack, David. 1965. Evolutionary Ecology. *J. Animal Ecology* 34:223–31.

Lamotte, Maxime. 1969. Representation synthetique des aspects statique et dynamique de la structure trophique d'un ecosysteme. *C. R. Acad. Sci.* Paris 268:2952–55.

―――. 1978. La savane preforestiere de Lamto, Côte d'Ivoire, pp. 231–311. In M. Lamotte and F. Bourliere (eds.), *Problemes d'ecologies*. Masson, Paris.

Lamotte, M., and F. Bourliere (eds.). 1978. *Problemes d'ecologie, structure et fonctionnement des ecosystemes terrestres*. Masson, Paris.

Larson, K. H. 1963. Continental, close-in fall out: Its history, measurement and characteristics, pp. 19–25. In V. Schultz and A. W. Klement, Jr. (eds.), *Radioecology*. Reinhold, New York.

Lavrin, Janko. 1946. *Tolstoy: An Approach*. Russell & Russell, New York.

LeFebvre, E. A. 1962. Energy metabolism in the pigeon Columba livia *at rest and in flight*. Ph.D. diss., Univ. of Minnesota, Minneapolis.

———. 1964. The use of $D_2^{18}O$ for measuring energy metabolism in *Columba livia* at rest and in flight. *Auk* 81:403–16.

Lefeuvre, J. C. 1990. La recherche en écologie en France. Heur et malheur d'une discipline en difficulté. *Courier de la Cellue Environnement del'INRA* 13:17–26.

Lemmel, Hans. 1939. *Die Organismusidee in Möllers Dauerwaldgedanken*. Verlag Jules Springer, Berlin.

Leopold, Aldo. 1949. *A Sand County Almanac and Sketches Here and There*. Oxford Univ. Press, London.

Levins, Richard, and Richard Lewontin. 1985. *The Dialectical Biologist*. Harvard Univ. Press. Cambridge.

Liebig, J. 1876. *Die Chemie in ihrer Anwendung auf Agrikultur und Physiologie*. 9th ed. (V. Zollner). Braunschweig.

Lieth, H. 1962. *Die Stoffproduktion der Pflanzendecke*. Gustav Fischer Verlag, Stuttgart.

Likens, Gene E. 1985. An experimental approach to the study of ecosystems. The Fifth Tansley Lecture. *J. Ecology* 73:381–96.

Likens, G. E., F. H. Bormann, N. M. Johnson, D. W. Fisher, and R. S. Pierce. 1970. The effect of forest cutting and herbicide treatment on nutrient budgets in the Hubbard Brook watershed ecosystem. *Ecol. Monographs* 40(1): 23–47.

Likens, G. E., and F. H. Bormann. 1974. Acid rain: A serious regional problem. *Science* 184(4143): 1176–79.

Likens, G. E., N. M. Johnson, J. N. Galloway, and F. H. Bormann. 1976. Acid precipitation: Strong and weak acids. *Science* 194(4265): 643–45.

Likens, G. E., F. H. Bormann, R. S. Pierce, J. S. Eaton, and N. M. Johnson. 1977. *Biogeochemistry of a Forested Ecosystem*. Springer-Verlag, New York.

Likens, G. E., E. F. Wright, J. N. Galloway, and T. J. Butler. 1979. Acid rain. *Scientific American* 241:43–51.

Lindeman, Raymond L. 1941a. *Ecological Dynamics in a Sensecent Lake*. Ph.D. diss., Univ. of Minnesota.

———. 1941b. Seasonal food-cycle dynamics in a senescent lake. *Am. Midland Naturalist* 26:636–73.

———. 1942. The trophic-dynamic aspect of ecology. *Ecology* 23(4): 399–418.

Lord, R. D., Jr., F. C. Bellrose, and W. W. Cochran. 1962. Radiotelemetry of the respiration of a flying duck. *Science* 137:39–40.

Lotka, Alfred J. [1925] 1956. *Elements in Mathematical Biology*. Dover, New York. (Originally published as *Elements of Physical Biology* by Williams and Wilkins, Baltimore.)

Lovejoy, Arthur O. 1936. *The Great Chain of Being*. Harvard Univ. Press, Cambridge.

Lugo, Ariel E. 1990. Fringe Wetlands, pp. 143–69. In Ariel E. Lugo, Mark Brinson, and Sandra Brown (eds.), *Forested Wetlands: Ecosystems of the World*, no. 15. Elsevier, Amsterdam.

Lugo, Ariel E., M. Sell, and S. C. Snedaker. 1976. Mangrove Ecosystem Analysis, pp. 113–45. In B. C. Patten (ed.), *Systems Analysis and Simulation in Ecology*, vol. 4. Academic Press, New York.

MacArthur, Robert. 1955. Fluctuations of animal populations and a measure of community stability. *Ecology* 36:533–36.

———. 1965. Patterns of species diversity. *Biol. Reviews* 40:510–33.

McAtee, W. L. 1907. Census of four square feet. *Science* 26:447–49.

Macfadyen, Amyan. 1949. The meaning of productivity in biological systems. *J. Animal Ecology* 17:75–80.

———. 1957. *Animal Ecology: Aims and Methods*. Pitman, London.

———. 1962. Soil arthropod sampling. *Advances in Ecological Research* 1:1–32.

McIntosh, Robert P. 1975. "Individualistic Ecologist, 1882–1975: His contributions to ecological theory," by H. A. Gleason. *Bull. Torrey Botanical Club* 102:253–73.

———. 1985. *The Background of Ecology: Concept and Theory*. Cambridge Univ. Press, Cambridge.

MacMahon, James A., Donald Phillips, James V. Robinson, and David J. Schimpf. 1978. Levels of biological organization: An organism-centered approach. *BioScience* 28(11): 700–704.

Maldague, M. 1984. The biosphere reserve concept: Its implementation and its potential as a tool for integrated development, pp. 376–401. In F. di Castri, F. W. G. Baker, and M. Hadley (eds.), *Ecology in Practice*, pt. 1. Tycooly International, Dublin.

Mann, K. H. 1969. The dynamics of aquatic ecosystems. *Advances in Ecological Research* 6:1–81.

Manning, W. M., C. Juday, and M. Wolf. 1938. Photosynthesis of aquatic plants at different depths in Trout Lake. *Trans. Wisc. Acad. Sci., Arts, Lett.* 31:377–410.

Mar, Brian. 1977. *Regional Environmental Systems: Assessment of RANN projects.* Dept. Civil Engineering, Univ. of Washington.

Margalef, D. Ramon. 1957. Information theory in ecology. *Memorias Real Acad. Ciencias y Artes Barcelona* 23:373–449.

———. 1962. Modelos fisicos simplificados de poblaciones de organismos. *Memorias Real Acad. Ciencias y Artes Barcelona* 34:83–146.

———. 1963. On certain unifying principles in ecology. *Am. Naturalist* 97:357–74.

Marsh, George Perkins. [1864] 1965. *Man and Nature.* Harvard Univ. Press, Cambridge, Mass.

Meyer, Judy, and Gene E. Likens. 1979. Transport and transformation of phosphorus in a forest stream ecosystem. *Ecology* 60(6): 1255–69.

Meyer-Abich, A. 1938. *Geleitwort zu Smuts "Die holistische Welt."* Berlin.

———. 1948. *Naturphilosophie auf neuen Wegen.* Hippokrates Verlag, Stuttgart.

Mitchell, Roger, Ramona A. Mayer, and Jerry Downhower. 1976. An evaluation of three biome programs. *Science* 192:859–65.

Möbius, Karl. 1877. *Die Auster und die Austernwirthschaft.* Verlag von Wiegandt, Hempel and Parey, Berlin. Trans. H. J. Rice and published in Fish and Fisheries Annual Report of the Commission for the year 1880, vol. 3, no. 29, app. H, pp. 681–747.

Möller, C. M. 1945. Untersuchungen über Laubmenge, Stoffverlust und Stoffproduktion des Waldes. *Forstl. Forsogsu. Danm.* 17:1–287.

Möller, C. M., D. Müller, and J. Nielsen. 1954. Ein Diagramm der Stoffproduktion in Buchenwald. *Ber. Schweiz. Bot. Ges.* 64:487–94.

Monsi, Masami. 1960. Dry-matter reproduction in plants 1. Schemata of dry-matter reproduction. *Botanical Magazine* (Tokyo) 73:81–90.

Monsi, M., Z. Uchijima, and T. Oikawa. 1973. Structure of foliage canopies and photosynthesis. *Ann. Rev. Ecol. Syst.* 4:301–27.

Mooney, Harold. 1975. Plant Physiological Ecology: A Synthetic View, pp. 19–36. In F. J. Vernberg (ed.), *Physiological Adaptations to the Environment.* Intext, New York.

Mori, S. (ed.). 1979. *Analysis of the Present State of Ecology and the Future Direction.* Tokyo.

Moser, W., and J. Peterson. 1982. Limits to Obergurgl's growth. *Ambio* 10(2/3): 68–72.

Müller, D. 1951. Analyse der Stoffproduktion von Gerste. *Die Bodenkultur* 5:129.

Murdoch, William W. 1966. "Community structure, population control, and competition"—A critique. *Am. Naturalist* 100:219–26.

Mysterud, Iver. 1971. *Forurensning og biologisk miljovern*. Universitetsforlaget, Oslo.

Naumann, Einar. 1932. *Grundzüge der regionalen Limnologie*. E. Schweizerbart'sche Verlagsbuchhandlung, Stuttgart.

Naumann, E., and A. Thienemann. 1922. Vorschlag zur Gründung einer internationalen Vereinigung für theoretische und angewandte Limnologie. *Arch. Hydrobiol.* 13:585–605.

Needham, James G., and J. T. Lloyd. 1916. *The Life of Inland Waters*. Comstock, Ithaca.

Numata, Makoto. 1990. The development of ecology in Japan. *Ecology International* 18:1–12.

Publications by Eugene P. Odum

1953. *Fundamentals of Ecology*. W. B. Saunders, Philadelphia.

1959. *Fundamentals of Ecology*. 2d ed. W. B. Saunders, Philadelphia.

1960. Organic production and turnover in old field succession. *Ecology* 41:34–49.

1963. *Ecology*. Holt Rinehart and Winston, New York.

1968. Historical review of the concepts of energy flow in ecosystems. *Am. Zool.* 8:11–18.

1969. The strategy of ecosystem development. *Science* 164:262–70.

1977. Ecology: The commonsense approach. *The Ecologist* 7(7): 250–53.

Odum, E. P., and J. D. Perkinson, Jr. 1951. Relation of lipid metabolism to migration in birds: Seasonal variation in body lipids of the migrating white-throated sparrow. *Physiol. Zool.* 24(3): 216–30.

Odum, E. P., and C. E. Connell. 1956. Lipid levels in migrating birds. *Science* 123:892–94.

Odum, E. P., and James C. Major. 1956. The effect of diet on photoperiod-induced lipid deposition in the white-throated sparrow. *The Condor* 58:222–28.

Odum, E. P., E. J. Kuenzler, and Sister N. X. Blunt. 1958. Uptake of P^{32} and primary productivity in marine benthic algae. *Limnology and Oceanography* 3(3): 340–45.

Odum, E. P., and Roger W. Bachman. 1960. Uptake of Zn^{65} and primary productivity in marine benthic algae. *Limnology and Oceanography* 5(4): 349–55.

Odum, E. P., C. E. Connell, and H. L. Stoddard. 1961. Flight energy and estimated flight ranges of some migrating birds. *Auk* 78:515–27.

Odum, E. P., and A. J. Pontin. 1961. Population density of the underground ant, *Lasius flavus*, as determined by tagging with P^{32}. *Ecology* 42(1): 186–88.

Odum, Howard T. 1956a. Primary production in flowing waters. *Limnology and Oceanography* 1:102–17.

———. 1956b. Efficiencies, size of organisms, and community structure. *Ecology* 37:592–97.

———. 1957. Trophic structure and productivity of Silver Springs, Florida. *Ecol. Monographs* 27:55–112.

———. 1960. Ecological potential and analogue circuits for the ecosystem. *Am. Scientist* 48:1–8.

———. 1971. *Environment, Power and Society*. Wiley-Interchange, New York.

Odum, Howard T., and James R. Johnson, Jr. 1955. Silver Springs . . . And the balanced aquarium controversy. *Science Counselor* (December).

Odum, Howard T., and Eugene P. Odum. 1955. Trophic structure and productivity of a windward coral reef community on Eniwetok atoll. *Ecol. Monograph* 25:291–320.

Odum, Howard T., and Richard C. Pinkerton. 1955. Time's speed regulator: The optimum efficiency for maximum power output in physical and biological systems. *Am. Scientist* 43:331–43.

Odum, Howard T., and Charles M. Hoskins. 1958. Comparative studies on the metabolism of marine waters. *Inst. of Marine Science, U. Texas* 5:16–46.

Odum, Howard T., John E. Cantlon, and Louis S. Kornicker. 1960. An organizational hierarchy postulate for the interpretation of species-individual distributions, species entropy, ecosystem evolution and the meaning of a species-variety index. *Ecology* 41:395–99.

Odum, Howard W. 1936. *Southern Regions of the United States*. Univ. of North Carolina Press, Chapel Hill.

Olson, J. S. 1965. Equations for Cesium transfer in a *Liriodendron* forest. *Health Physics* 11:1385–92.

O'Neill, R. V., D. L. DeAngelis, J. B. Waide, and T. F. H. Allen. 1986. *A Hierarchical Concept of Ecosystems*. Monogr. Population Biol 23. Princeton Univ. Press, Princeton.

Oosting, H. J. 1942. An ecological analysis of the plant communities of Piedmont, North Carolina. *Am. Midland Naturalist* 28:1–126.

Osborn, Fairfield. 1948. *Our Plundered Planet*. Little Brown, Boston.

———. 1953. *Limits of the Earth*. Little Brown, Boston.

Ovington, J. D. 1959a. The calcium and magnesium contents of tree species grown in close stands. *The New Phytologist* 58:164–75.

————. 1959b. The circulation of minerals in plantations of *Pinus sylvestris L. Annals of Botany* 23:229–39.

————. 1962. Quantitative ecology and the woodland ecosystem. *Advances in Ecological Research* 1:103–92.

Paine, Robert T. 1964. Ash and calorie determinations of sponge and opisthobranch tissues. *Ecology* 45:384–87.

————. 1969. A notes on trophic complexity and community stability. *Am. Naturalist* 103:91–93.

Patten, Bernard C. 1959. An introduction to the cybernetics of the ecosystem: The trophic-dynamic aspect. *Ecology* 40:221–31.

————. 1972. A simulation of the shortgrass prairie ecosystem. *Simulation* 19:177–86.

Patten, Bernard, and Eugene P. Odum. 1981. The cybernetic nature of ecosystems. *Am. Naturalist* 118:886–95.

Peters, Sir Rudolph. 1975. Statement, p. 3. In E. B. Worthington (ed.), *The Evolution of* IBP. Cambridge Univ. Press, Cambridge.

————. 1976. Tautology in evolution and ecology. *Am. Naturalist* 110:1–12.

Phillips, John. 1931. The biotic community. *J. Ecol.* 19:1–24.

————. 1934. Succession, development, the climax, and the complex organism: An analysis of concepts. Pt. 1. *J. Ecol.* 22:554–71.

————. 1935a. Succession, development, the climax, and the complex organism: An analysis of concepts. Pt. 2. Development and the climax. *J. Ecol.* 23:210–43.

————. 1935b. Succession, development, the climax, and the complex organism: An analysis of concepts. Pt. 3. The complex organism: Conclusion. *J. Ecol.* 23: 488–508.

————. 1954. A tribute to Frederic E. Clements and his concepts in ecology. *Ecology* 35(2): 114–15.

Phillipson, John. 1966. *Ecological Energetics*. Edward Arnold, London.

Pimm, S. L. 1982. *Food Webs*. Chapman and Hall, London.

Popper, Karl. 1959. *The Logic of Scientific Discovery*. Harper and Row, New York.

Porter, Warren P., and David M. Gates. 1969. Thermodynamic equilibria of animals with environment. *Ecol. Monographs* 39:245–70.

Pound, Roscoe, and Frederic E. Clements. 1897. *The Phytogeography of Nebraska*. Pt. 1, General Survey. 2d ed. Univ of Nebraska, Lincoln.

Preston, F. W. 1948. The commonness and rarity of species. *Ecology* 29:254–83.

———. 1960. Time and space and the variation of species. *Ecology* 41:785–90.

———. 1962. The canonical distribution of commonness and rarity. *Ecology* 43:185–215.

Rackham, Oliver. 1980. *The Ancient Woodlands of England*. Edward Arnold, London.

Rashevsky, N. 1956. Topology and life: In search of general mathematical principles in biology and sociology. *General Systems* 1:123–38.

Rawson, D. S. 1930. *The Bottom Fauna of Lake Simcoe and Its Role in the Ecology of the Lake*. Univ. Toronto Stud., Publ. Ontario Fish Res. Lab., no. 40.

Reif, Charles B. 1986. Memories of Raymond Laurel Lindeman. *Bull. Ecol. Soc. Am.* 67:20–25.

Ricker, William E. 1946. Production and utilization of fish populations. *Ecol. Monographs* 16:374–91.

Riley, Gordon A. 1941. Plankton studies: Long Island Sound. *Bull. Bingham Oceanogr. Coll.* 7:1–89.

———. 1944. The carbon metabolism and photosynthetic efficiency of the earth as a whole. *Am. Scientist* 32:129–34.

Risser, Paul G., E. C. Birney, H. D. Blocker, S. W. May, W. J. Parton, and J. A. Wiens. 1981. *The True Prairie Ecosystem*. Hutchinson and Ross, Stroudsburg.

Rodhe, H. 1972. A study of the sulfur budget for the atmosphere over northern Europe. *Tellus* 24:128–38.

Rodhe, Wilhelm. 1974. The International Association of Limnology: Creation and function. *Mitt. Int. Verein. Limnol.* 20:44–70.

Russett, Cynthia Eagle. 1966. *The Concept of Equilibrium in American Social Thought*. Yale Univ. Press, New Haven.

Ruttner, F. 1953. *Fundamentals of Limnology*. Trans. D. G. Frey and F. E. J. Fry. Univ. Toronto Press.

Ryszkowski, Lech, and S. Balazy. 1988. Environmental hazards in Poland, and a strategy for conservation of living natural resources. *Ecology International* 16:17–28.

Saarinen, Esa. 1980. *Conceptual Issues in Ecology*. D. Reidel, Dordrecht.

Saeki, Toshiro. 1959. Variation of photosynthetic activity with aging leaves and total photosynthesis in a plant community. *Botanical Magazine* (Tokyo) 72:404–8.

Saeki, Toshiro, and Nobuo Nomoto. 1958. On the seasonal change of photosynthetic activity of some deciduous and evergreen broadleaf trees. *Botanical Magazine* (Tokyo) 71:235–41.

Salm, Peter. 1971. *The Poem as Plant: A Biological View of Goethe's Faust*. Case Western Reserve Univ. Press, Cleveland.

Sargent, M. C., and T. S. Austin. 1949. Organic productivity of an atoll. *Am. Geophysical Union Trans.* 30(2): 245–49.

Schindler, D. W., K. H. Mills, D. F. Malley, D. L. Findlay, J. A. Sheaver, I. J. Davies, M. A. Turner, G. A. Lindsey, and D. R. Cruikshank. 1985. Long-term ecosystem stress: The effects of years of experimental acidification on a small lake. *Science* 228:1395–1401.

Schmidt, W. 1983. *Verhandlungen.* vol. 11. Gesellschaft für Ökologie. Festschrift für Heinz Ellenberg. Göttingen.

Scott, D. 1965. The determination and use of thermodynamic data in ecology. *Ecol.* 46(5): 673–80.

Shannon, C. E., and W. Weaver. 1949. *The Mathematical Theory of Communication.* Univ. of Illinois Press, Urbana.

Sheail, John. 1987. *Seventy-Five Years in Ecology.* The British Ecological Society. Blackwell, Oxford.

Shelford, Victor E. 1913. *Animal Communities in Temperate America as Illustrated in the Chicago Region: A Study in Animal Ecology.* Univ. of Chicago Press, Chicago.

———. 1918. Conditions of existence, pp. 21–60. In H. Ward and G. Whipple (eds.), *Freshwater Biology.* Wiley, New York.

———. 1926. *A Naturalist's Guide to the Americas.* Williams and Wilkins, Baltimore.

———. 1932. Basic principles on the classification of communities and habitats and the use of terms. *Ecology* 13:105–20.

Simberloff, Daniel. 1980. A succession of paradigms in ecology: Essentialism to materialism and probabilism, pp. 63–99. In E. Saarinen (ed.), *Conceptual Issues in Ecology.* D. Reidel, Dordrecht.

Sioli, H. 1957. Beiträge zur regionalen Limnologie des Amazonasgebietes. Chap. 4, Limnologische Untersuchungen in der Region Belem-Braganca ("Zona Brigantina") im Staate Para, Brazil. *Arch. Hydrobiol.* 53(2): 161–222.

———. 1963. Beiträge zur regionalen Limnologie des Amazonasgebietes. V. Die Gewässer der Karbonstreifen Unteramazoniens (sowie einige Angaben über Gewässer der anschliessenden Devonstreifen). *Arch. Hydrobiol.* 59:311–50.

———. 1968. Zur Ökologie des Amazonasgebietes, pp. 137–70. In *Biogeography and Ecology in South America.* Vol. 1. Monographiae Biologicae 18. W. Junk, The Hague.

Slobodkin, L. Basil. 1961. Preliminary ideas for a predictive theory of ecology. *Am. Naturalist* 95:147–53.

———. 1962. Energy in Animal Ecology, pp. 69–101. In J. B. Cragg (ed.), *Advances in Ecological Research* 1. Academic Press, London.

———. 1964. The strategy of evolution. *Am. Scientist* 52:342–57.

Slobodkin, L. B., and S. Richman. 1961. Calories/gm in species of animals. *Nature* 191:299.

Slobodkin, L. B., F. E. Smith, and N. G. Hairston. 1967. Regulation in terrestrial ecosystems and the implied balance of nature. *Am. Naturalist* 101:109–24.

Smuts, Jan C. 1926. *Holism and Evolution*. Macmillan, New York.

Soderquist, Thomas. 1986. *The Ecologists: From Merry Naturalists to Saviours of the Nation*. Almquist and Wiksell, Stockholm.

Spencer, Herbert. 1896. *The Principles of Biology*. 2 vols. D. Appleton, New York.

Spiegel, H. W. 1971. *Growth of Economic Thought*. Prentice Hall, Englewood Cliffs.

Sprugel, George, Jr. 1975. John N. Wolfe, 1910–1974. *Bull. Ecol. Soc. Am.* 56(3): 20–22.

Strom, K. M. 1928. Recent Advances in Limnology. *Proceedings Linnean Society London* 1928:96–110.

Sukachev, V. 1960. Relationship of biogeocenosis, ecosystem and facies. *Soviet Soil Science* 6:579–84. (English trans.).

Sukachev, V., and N. Dylis. 1964. *Fundamentals of Forest Biogeocoenology*. Trans., pub. 1969. Oliver and Boyd, Edinburgh.

Summerhayes, V. S., and C. S. Elton. 1923. Contributions to the ecology of Spitzbergen and Bear Island. *J. Ecology* 11:214–86.

Swank, Wayne T. 1988. Stream chemistry responses to disturbances, pp. 339–58. In Swank and Crossley, *Forest Hydrology and Ecology at Coweeta*.

Swank, Wayne T., and D. A. Crossley, Jr. (eds.) 1988. *Forest Hydrology and Ecology at Coweeta*. Springer-Verlag, New York.

Publications by Arthur G. Tansley

1911. *Types of British Vegetation*. Cambridge Univ. Press, Cambridge.

1920. *The New Psychology and Its Relation to Life*. Allen and Unwin, London.

1922. *Elements of Plant Biology*. Allen and Unwin, London.

1929. *Succession: The Concept and Its Values*. Proc. Int. Congress of Plant Science, Ithaca. 1:677–86.

1935. *The use and abuse of vegetational concepts and terms*. Ecology 16(3): 284–307.

1939a. *The British Islands and Their Vegetation*. Cambridge at the University Press, Cambridge.

1939b. British ecology during the past quarter-century: The plant community and the ecosystem. *J. Ecol.* 27:513–30.

1942. *The Values of Science to Humanity.* Herbert Spencer Lecture. Allen and Unwin, London.

1947. The early history of plant ecology in Britain. *J. Ecology* 35:130–37.

1949. *The British Islands and Their Vegetation.* 2 vols. Cambridge University Press.

1952. *Mind and Life: An Essay in Simplification.* Allen and Unwin, London.

Tansley, A. G., and T. F. Chipp. 1926. *Aims and Methods in the Study of Vegetation.* British Empire Vegetation Committee and the Crown Agents for the Colonies, London.

Taylor, Peter L. 1988. Technological optimism, H. T. Odum, and the partial transformation of ecological metaphor after World War II. *J. Hist. Biol.* 21(2): 213–44.

Teal, John M. 1957. Community metabolism in a temperate cold spring. *Ecol. Monographs* 27:283–302.

———. 1962. Energy flow in the salt marsh ecosystem of Georgia. *Ecology* 43:614–24.

Tester, J. 1963. Techniques for studying movements of vertebrates in the field, pp. 445–50. In V. Schultz and A. W. Klement, Jr. (eds.), *Radioecology.* Reinhold, New York.

Thienemann, August. 1918. Lebensgemeinschaft und Lebensraum. *Naturw. Wochenschrift N. F.* 17:282–90, 297–303.

———. 1925. *Die Binnengewässer Mitteleuropas.* E. Schweizerbart'sche Verlagsbuchhandlung, Stuttgart.

———. 1926. Der Nahrungskreislauf im Wasser. *Verh. deutsch. Zool. Ges.* 31:29–79.

———. 1931. Der Produktionsbegriff in der Biologie. *Arch. Hydrobiol.* 22:616–22.

———. 1939. Grundzüge einer allgemeinen Ökologie. *Arch. Hydrobiol.* 35:267–85.

Tilly, Laurence J. 1968. The structure and dynamics of Cone spring. *Ecol. Monographs* 38:169–97.

Tobey, Ronald C. 1981. *Saving the Prairies: The Life Cycle of the Founding School of American Plant Ecology, 1895–1955.* Univ. California Press, Berkeley, Los Angeles.

Todes, Daniel P. 1989. *Darwin without Mathus: The Struggle for Existence in Russian Evolutionary Thought.* Oxford Univ. Press, New York.

Troll, C. 1971. Landscape ecology (Geoecology) and Biogeocoenology: A terminological study. *Geoforum* 8:43–46.

Trommer, Gerhard. 1990. *Natur im Kopf: Die Geschichte ökologisch bedeutsamer Naturvorstellungen in deutschen Bildungskonzepten.* Deutscher Studienverlag, Weinheim.

Tucker, C. J., T. R. G. Townshend, and T. Goff. 1985. African land-cover classification using satellite data. *Science* 227:369–75.

Van Dyne, George M. 1966. *Ecosystems, Systems Ecology, and Systems Ecologists.* Oak Ridge Nat. Lab 3957.

———. 1972. Organization and management of integrated research, pp. 111–72. In J. N. R. Jeffers (ed.), *Mathematical Models in Ecology.* Blackwell, Oxford.

———. 1978. Forward, perspectives on the ELM model and modeling efforts, pp. v–xx. In G. S. Innis (ed.), *Grassland Simulation Model.* Springer-Verlag, New York.

———. 1980. Reflections and projections, pp. 881–921. In A. I. Breymeyer and G. M. Van Dyne (eds.), *Grasslands, Systems Analysis and Man.* Cambridge Univ. Press, Cambridge.

Vernadsky, Vladimir I. 1926. *Biosfera.* Nauerekhizdat, Leningrad.

———. 1929. *La biosphère.* Félix Alcan, Paris. (Abridged ed., *The Biosphere*, was published in 1986 by Synergetic Press, Oracle, Arizona.)

Vestel, Arthur G. 1914. Internal relations of terrestrial associations. *Am. Naturalist* 48:413–45.

Vogt, William. 1948. *Road to Survival.* William Sloane, New York.

Waddington, C. H. 1975. The origin, pp. 4–12. In E. B. Worthington, *The Evolution of IBP.* Cambridge Univ. Press, Cambridge.

Ward, Henry, and George C. Whipple. 1918. *Freshwater Biology.* Wiley, New York.

Ward, Lester F. 1897. *Dynamic Sociology or Applied Social Science.* 2 vols. D. Appleton, New York.

Wasmund, E. 1930. Lakustrische Unterwasserböden. *Handbuch der Bodenlehre* 5:97–190.

Watt, K. E. F. 1962. Use of mathematics in population ecology. *Ann. Rev. Ent.* 7:243–60.

Weber, Hermann. 1939a. Zur Fassung und Gliederung eines allgemeinen biologischen Umweltbegriffes. *Die Naturwissenschaften* 27(38): 633–44.

———. 1939b. Der Umweltbegriff der Biologie und seine Anwendung. *Der Biologe* 1939(7/8): 245–61.

Weiner, D. R. 1988. *Models of Nature, Ecology, Conservation and Cultural Revolution in Soviet Russia.* Indiana Univ. Press, Bloomington.

Westlake, D. F. 1963. Comparisons of plant productivity. *Biol. Rev.* 38:385–425.

Wheelwright, Philip. 1959. *Heraclitus.* Princeton Univ. Press, Princeton.

White, Gilbert. [1789] 1981. *The Illustrated Natural History of Selborn*. Macmillan, London.

Whitehead, Alfred North. 1944. *Science and the Modern World*. Macmillan, New York.

Whittaker, Robert H. 1953. A consideration of climax theory: The climax as a population and pattern. *Ecol. Monographs* 23:41–78.

Wiegert, Richard G. 1964. Population energetics of meadow spittlebugs (*Philaenus spumarius* L.) as affected by migration and habitat. *Ecol. Monographs* 34:217–41.

———. 1965. Intraspecific variation in calories/g of meadow spittlebugs (*Philaenus spumarius* L.) *Bioscience* 15:543–45.

———. 1968. Thermodynamic considerations in animal nutrition. *Am. Zool.* 8:71–81.

———. 1976. *Ecological Energetics*. Benchmark Papers in Ecology 4. Dowden, Hutchinson and Ross, Stroudsburg.

———. 1988. Holism and reductionism in ecology: Hypotheses, scale, and system models. *Oikos* 53(2): 267–69.

Wiegert, R. G., and D. F. Owen. 1971. Trophic structure, available resources and population density in terrestrial vs. aquatic ecosystems. *J. Theoretical Biology* 30:69–81.

Wiener, Norbert. 1948. *Cybernetics*. Wiley, New York.

Williams, C. B. 1944. Some applications of the logarithmic series and the index of diversity to ecological problems. *J. Ecology* 32:1–44.

———. 1953. The relative abundance of different species in a wild animal population. *J. Animal Ecology* 22:14–31.

Wolfe, J. N. 1967. Radioecology: Retrospection and future, pp. xi–xii. In D. J. Nelson and F. C. Evans (eds.), *Symposium on Radioecology*. AEC Biol. & Medicine Conf. 670503.

Woltereck, Richard. 1940. *Ontologie des Lebendigen*. Ferdinand Enke Verlag, Stuttgart.

Wood, J. E., and E. P. Odum. 1964. A nine-year history of furbearer populations on the AEC Savannah River Plant area. *J. Mammalogy* 45(4): 540–51.

Woodmansee, R. G. 1978. Critique and analyses of the grassland ecosystem model ELM, pp. 257–81. In G. S. Innis (ed.), *Grassland Simulation Model*. Springer-Verlag, New York.

Woodwell, G. M., and H. H. Smith (eds.). 1969. *Diversity and stability in ecological systems*. Brookhaven Symp. Biol., no. 22. Brookhaven Nat. Lab., AEC.

Worster, Donald. 1977. *Natures Economy: The Roots of Ecology*. Sierra Club Books, San Francisco.

Worthington, E. B. (ed.). 1975. *The evolution of IBP*. Cambridge Univ. Press, Cambridge.

Wynne-Edwards, V. C. 1962. *Animals Dispersion in Relation to Social Behavior*. Oliver and Boyd, Edinburgh.

Yeo, Richard R. 1986. Scientific method and the rhetoric of science in Britain, 1830–1917, pp. 259–97. In J. A. Schuster and R. R. Yeo (eds.), *The Politics and Rhetoric of Scientific Method*. D. Reidel, Dordrecht.

Zeuthan, E. 1953. Oxygen uptake and body size in organisms. *Quarterly Review of Biology* 28:1–12.

INDEX